THE OPEN UNIVERSITY
Science: A Second Level Course

BLOCK 1 EARTH COMPOSITION
elements, minerals and rocks

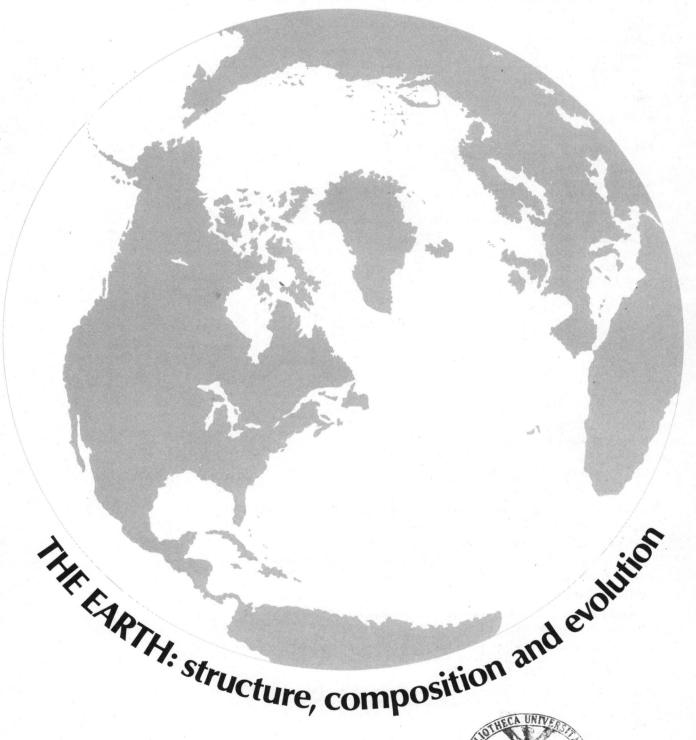

THE EARTH: structure, composition and evolution

THE OPEN UNIVERSITY PRESS

The S237 Course Team

Chairman

Geoff Brown

Course Coordinator

Angela Colling

Authors

Geoff Brown
Steve Drury
Andrew Fleet (*Goldsmiths' College*)
Peter Francis
Chris Hawkesworth
Julian Pearce
Peter Smith
Sandra Smith
Richard Thorpe
John Wright

Editor

Perry Morley

Other Members

Nick Brenton (*BBC*)
Stuart Carter (*BBC*)
Alan Cooper (*Physics*)
Laurie Melton (*Librarian*)
Fran Page (*Designer*)
Eileen Scanlon (*IET*)
John Simmons (*BBC*)
Barrie Whatley (*BBC*)
Geoff Yarwood (*Staff Tutor*)

8503294 – QE

The Open University Press
Walton Hall, Milton Keynes
MK7 6AA

First published 1981. Reprinted 1982

Designed by the Graphic Design Group of the Open University.

Typeset by Technical Filmsetters Europe Ltd., 76 Great Bridgewater Street, Manchester M1 5JY and printed in Great Britain by Staples Printers, St Albans Limited at The Priory Press.

ISBN 0 335 16055 7

This text forms part of an Open University course. The complete list of Blocks in the course appears at the end of this text.

For general availability of supporting material referred to in this text, please write to Open University Educational Enterprises Limited, 12 Cofferidge Close, Stony Stratford, Milton Keynes, MK11 1BY, Great Britain.

Further information on Open University courses may be obtained from the Admissions Office, The Open University, P.O. Box 48, Walton Hall, Milton Keynes, MK7 6AB.

1.2

Contents

Table A

List of scientific terms and concepts used in this Block

Introduced in the Science Foundation Course	S100* Unit No.	S101** Unit No.	Introduced in the Science Foundation Course	S100* Unit No.	S101** Unit No.
age of the Earth	27	26	mass spectrometry	6	10/11
andesite	26, UTE†	27	mean value	HED	HED
atmospheric evolution	27	28	metallic bonding	8	13
atomic number	6	10/11	metamorphism	24	27
basalt	22	4	metamorphic rocks	24	27
calcareous ooze	24	27	meteorites	27	28
chemical equilibrium	9	14	mineral	24	27
chemical weathering	24	27	momentum, conservation of	3	29
constructive interference	28	9	mudstone	UTE	4
constructive plate boundaries	25	6/7	nebular theory	27	28
contact metamorphism	UTE	27	nuclear fission	31	30
continuous spectrum	6	10/11	nuclear fusion	31	30
core (of the Earth)	22	4, 28	oceanic lithosphere	25	6/7
covalent bonding	8, 10	13	oxidation state	8	28
crust (of the Earth)	22	4, 28	oxygen sharing	UTE	27
density	3	3	peridotite	22	4
destructive interference	28	9	Periodic Table	7, 8	13
destructive plate boundaries	25	6/7	pH scale	9	14
electronegativity	8	13	physical weathering	24	27
electron sub-shells	6, 7	10/11	planet formation	UTE	28
endothermic reactions	11	15	planetary layering	UTE	28
energy of environment	26	27	Planck's constant	6	9
enthalpy of reaction	11	15	plate tectonics	25	6/7
evaporites	25	27	radioactive decay	6, 31	26, 30
exothermic reactions	11	15	regional metamorphism	26	27
Fraunhofer lines	6	10/11	rock cycle	24	27
frequency distribution curve	HED	HED	rock dating	26	26
gabbro	UTE	4	sandstone	UTE	4
gneiss	UTE	4	schist	UTE	6/7
granite	26	4	sedimentary rocks	22, 24	27
greenhouse effect	UTE	28, 32	sedimentation	24	27
hypothesis	1	1	seismology	22	4
igneous rocks	22, 24	27	slate	UTE	27
ionic bonding	8	13	silicate minerals	24	27
Le Chatelier's principle	9	14	solar spectroscopy	6	10/11
limestone	26	4	Solar System	22, 27	1, 28
magma	24	4	standard deviation	HED	HED
mantle (of the Earth)	22	4, 28	unbinding energy	31	30
mass number	6	10/11			

* The Open University (1971) S100 *Science: A Foundation Course*, The Open University Press.
** The Open University (1979) S101 *Science: A Foundation Course*, The Open University Press.
† S100 Set book: Gass, I. G., Smith, P. J. and Wilson, R. C. L. (eds) (1972) *Understanding the Earth*, 2nd edn, Artemis Press.

Introduced or developed in this Block	Page No.	Introduced or developed in this Block	Page No.
absorption spectra	14	amphiboles	54
accuracy	86	amphibolite	66
achondrites	12	asteroids	11
acid rocks	68	atmophile elements	32
alkali feldspars	58	basic rocks	68
Allende meteorite	26	biotite	56
aluminosilicate minerals	80	Bragg's equation	40

1 Study guide

This Course is about the solid Earth. Its principal aim is to describe the nature of our planet, its physical and chemical infrastructure and its evolution. As with most branches of modern science you will find that, to achieve this aim, we must call on evidence from a wide variety of sources. Two of the most important corner-stones of this Course that have led to many new and exciting discoveries in recent decades are the hybrid sub-disciplines: *geophysics* and *geochemistry*. For example, you may already realize that during the 1960s the high quality magnetometer evidence of *magnetic 'stripes'* on the ocean floor led to the confirmation of sea-floor spreading and so was a major geophysical factor in the development of the theory of plate tectonics. Some of the new discoveries in geophysics that you will find in this Course include the detailed characteristics of earthquakes and the variation of physical properties, such as density, through the Earth's interior. These are discussed, with the techniques that made these discoveries possible, in Block 2. At the end of Block 2 you will find a preliminary description of the composition of the Earth's layered interior, and in Block 3 we shall be using the results of geochemical techniques to describe the processes of melting and crystallization that determine these compositions. The knowledge which you will have from Blocks 1–3 will allow us to discuss the origin of the Earth's internal temperature and the convection of its deep interior (Block 4). Block 5 discusses the geochemistry of sedimentary processes; Block 6 looks at both the geochemistry and geophysics of hydrocarbon and metalliferous mineral deposits. Finally, Block 7 reviews the entire history of the Earth, from its formation 4 600 million years ago, to the present day, concentrating on the interpretation of the present geological features of the crust in terms of past internal and surface processes.

This rapid summary of the Course is developed in more detail in the *Introduction and Guide* to S237. We have included a synopsis here to show you that geophysics and geochemistry are intimately linked in most of the aspects of the Earth into which we shall be enquiring. Geophysics is concerned primarily with the physical properties of the Earth, particularly the deep and inaccessible parts which we are unable to examine directly. It uses observable properties such as density, magnetism and electrical conductivity as constraints in the determination of the nature and behaviour of those deep layers. This has become a much more important part of the Earth sciences in the last thirty years because of technological innovation. In contrast, the more mature sub-discipline of geochemistry has developed progressively during the past century. It is concerned with the analysis and distribution of the chemical elements in different parts of the Earth and with the partly physical and partly chemical processes of *melting, metamorphism*[A]* and *sedimentation*[A].

Not surprisingly, it would be difficult to cover properly all the topics embraced by geophysics and geochemistry in a half-credit course. So our policy has been to use examples that illustrate some of the basic principles and processes as clearly as possible, so that you will appreciate the flavour of these fascinating areas of science. Because it impinges on disciplines other than those of the Earth sciences, this Course draws on different aspects of your Foundation Course, and so one of the main aims of this Block is to remind you about the most relevant parts of your earlier studies and to show how they are related in the context of this Course.

First, therefore, in Section 2 we shall remind you about the position of the Earth in the *Solar System*[A] and its relationship to other planets and to different kinds of *meteorites*[A]. We shall introduce the chemical composition of the dominant body of the Solar System—the Sun—using *solar spectroscopy*[A] and shall consider the ways in which all the chemical elements now present may have been produced inside stars by *nuclear fusion*[A] processes. Then Section 2 will outline the way in which these elements may have been distributed in the Solar System during *planet formation*[A] and also how they are distributed inside the Earth today. TV 01 complements this Section of the Block.

In Section 3 we aim to provide some of the reasons for the distribution of the elements in the Earth. We shall start by revising the Periodic Table of the chemical elements, showing how their behaviour in natural systems is a function of their

---

* In this Block we have indicated terms and concepts that are re-used from the Science Foundation Course by a superscript A. You will find a list of these terms and concepts, with the appropriate references, in the first part of Table A.

properties. The geochemical classification of the elements is aided by considering *ionic*[A], *covalent*[A] and *metallic bonding*[A] in a natural context. The next step is to show how the crystalline structures of silicate minerals may be built up by different combinations of silicon and other geologically important 'major' elements with oxygen. The internal atomic structures and compositions of the different *silicate minerals*[A] determine their properties and, in turn, the silicates determine the physical and chemical features of rocks, the vast majority of which are made up of these minerals.

In Section 4 we shall look at the formation and chemical composition of rocks and will show you how the *rock cycle*[A] may equally be thought of in terms of many mutually interacting 'geochemical cycles'.

In these various ways Block 1 is an introduction to the Course, providing you with both a general view of the Earth and its formation, and a broad appreciation of the physical and chemical features of rocks and minerals. Other parts of the Course will elaborate on many of the subjects introduced here. To help you in planning your work, you may like to bear in mind our estimates of study time for each major Section in terms of Course-Unit equivalents (CUE):

Section 2	0.6 CUE
Section 3	1.2 CUE
Section 4	0.7 CUE

These estimates include an allowance for the two audiovision–Home Experiment linked sequences and the two television programmes associated with this Block. The first audiovision sequence (AV 01) looks at the internal atomic structures of silicate minerals and makes use of the ball-and-spoke models in your Home Experiment Kit; it should be completed at the beginning of Section 3.3. Before you start Section 4, you will need to study AV 02, which introduces the recognition of minerals in your Home Experiment Kit rocks and the origins of these rocks. Each audiovision sequence will require about one hour of study time. Apart from introducing the Course as a whole, TV 01 is used to illustrate Section 2 of this text whereas TV 03 will provide a link between minerals and rocks (Sections 3 and 4 of Block 1) and crustal structure (Block 2). (TV 02 will be used to introduce the earthquake section in Block 2.)

One or two more points about the format we have adopted in the Main Texts of S237: particularly important concepts and summaries are highlighted by the use of red rules in the text and there are also summaries of a more general nature at the end of each major text Section. Following each summary you will find that the Objectives for the appropriate Section are given together with some SAQs. To help you in your revision, all the Objectives are collected together at the end of each Block where they are cross-referenced to the relevant SAQs and ITQs. You should also note that the Main Texts are complemented by *Colour-plate Booklets* which are needed for most of the audiovision sequences. Finally, please do not be dismayed by the number of back-references to the Science Foundation Course: you will find the appropriate source Units listed in Table A of each Block. You will also find that most of these terms are clearly re-defined in S237.

A word about Units: In this Course we shall use the powers-of-ten notation which should be familiar to you from the Science Foundation Course. The standard abbreviations are shown in the margin. You will encounter a range of these: for example, the age of the Earth is measured in numbers of Ga (10^9 years) and the radius of an atom is measured in numbers of pm (10^{-12} m).

Prefix*	Symbol	Power of 10
tera	T	10^{12}
giga	G	10^9
mega	M	10^6
kilo	k	10^3
centi	c	10^{-2}
milli	m	10^{-3}
micro	μ	10^{-6}
nano	n	10^{-9}
pico	p	10^{-12}
femto	f	10^{-15}

* Note that when a prefix is placed in front of a unit, it in effect produces a new unit. Consequently, nm^2 (for instance) should be read as $(nanometres)^2$ and *not* as nano $\times (metres)^2$.

2 The planet Earth in a universal context

Study comment In the first major Section of this Block we look at the main features of the Solar System and introduce measurements of solar spectra that establish the composition of its dominant body: the Sun. The origin and relative abundances of the chemical elements in the Solar System (and elsewhere in the Universe) is related to nuclear fusion processes that take place inside stars (Section 2.2.2), processes that generated the material of which the Earth itself is also made. Section 2.3 gives a broad summary of the Earth's formation and segregation into a layered internal structure and so forms a prelude to later Blocks.

In a Course essentially about the Earth, one of the major aims is to account for the origin of our planet. We cannot do this by reference to the Earth alone, partly because we can examine directly only its outermost few kilometres and partly because the Earth is just one member of a single Solar System, consisting of the Sun and nine planets, whose components probably are intimately related by a common origin. By studying the Solar System as a whole, we can estimate its original composition and determine the physical and chemical processes that have probably led to its present state; this is the topic we shall now consider.

2.1 The Sun and the Solar System

Our Solar System is dominated by a single star—the Sun—which has a radius of $c.$ 7×10^5 km and a mass of $c.$ 2×10^{30} kg: it is of average size and mass for all stars. The density of the Sun ($c.$ 1.4×10^3 kg m^{-3}) is low, like that of most other stars, because it is composed predominantly of light gases such as hydrogen and helium. This deduction is supported by spectral observations, as you will see shortly. There is nothing special about the Sun: it is just one of about 10^{11} stars, some larger and some smaller than the Sun, which make up our *Galaxy*, a flattened spiral star structure (*CB**, Plate 1). The Sun is situated in one of the spiral arms of our Galaxy and, when you look into the 'Milky Way' at night, you can see the dense cluster of stars forming the spiral arm of which the Sun is one member. In the Universe, there are many thousands of other galaxies, some visible and others known to us only because of electromagnetic emissions other than light (such as X-rays). Some have been recorded by astronomers at distances up to 4.5×10^{22} km away from our Galaxy. Since electromagnetic radiation travels at 3×10^8 m s^{-1} (the speed of light) this means that light, or other radiation, from such distant galaxies has taken nearly 5 Ga (5×10^9 years) to reach us. In other words, since the *age of the Earth*[A] and the Solar System is $c.$ 4.6 Ga, it started on its way to the Earth before the Earth and the Solar System came into being! This is just one of many pieces of evidence to indicate that different stars are of different ages and, indeed, that stars are being born and are dying all the time (see also Section 2.2).

Galaxy

TABLE 1   Planetary data

	Mean distance from Sun (AU*)	Orbital period (Earth days or years)	Period of rotation on its axis (Earth days)	Radius relative to Earth (= 6 378 km)	Mass relative to Earth (= 6×10^{24} kg)	Mean density (10^3 kg m^{-3})	Number of satellites
Sun			25.4	109	343 000	1.4	
Mercury	0.39	88 days	59	0.38	0.055	5.42	0
Venus	0.72	225 days	−243	0.95	0.815	5.27	0
Earth	1.00	365 days	1.00	1.00	1.00	5.52	1
Mars	1.52	1.88 years	1.03	0.53	0.108	3.95	2
Jupiter	5.20	11.9 years	0.41	11.2	318	1.33	14
Saturn	9.52	29.5 years	0.43	9.5	95	0.69	10
Uranus	19.2	84.0 years	0.89	3.7	14.6	1.2	5
Neptune	30.1	164 years	0.53	3.9	17.2	1.7	2
Pluto	39.4	247 years	6.4	~0.5	0.1	~1.6	1
Moon	1.00	27.3 days	27.3	0.27	0.012	3.33	

* 1 AU (Astronomical Unit) is the mean distance of the Earth from the Sun: 1.496×10^8 km.

The Sun, one of perhaps 10^{15} stars in the Universe**, is a tiny and insignificant component of it, and the Earth is one of just four *inner* or *terrestrial planets* that orbit the Sun (see Table 1 and Figure 1). These inner planets are relatively small and dense compared to the four *outer* or *major planets* (not including Pluto) which are large and light. As you can see from Table 1, the Sun has a volume and mass 1 000 times greater than those of the largest planet, Jupiter (remember that volume is proportional to the cube of the radius). In turn, Jupiter is about 1 000 times larger than the Earth by volume, but only 300 times more massive. This is also reflected in the density column of Table 1 (*density*[A] is mass per unit volume).

inner planets

major planets

* The Open University (1981) S237 *Colour-plate Booklets*. The Open University Press.

**There are about 10^{15} stars in the Universe, with an average of 10^{11} stars per galaxy, and about 10^4 galaxies.

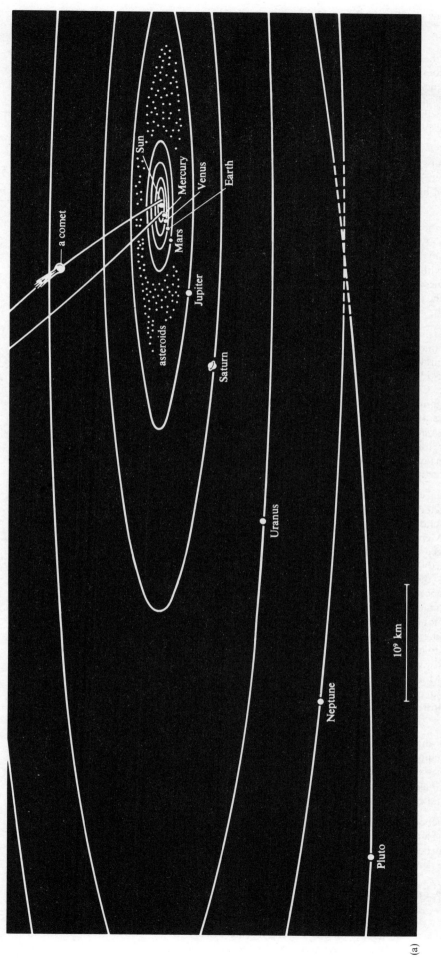

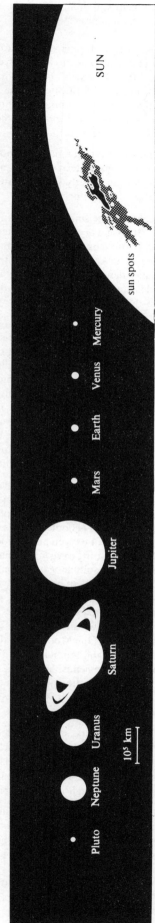

FIGURE 1 Summary of (a) the orbital paths and (b) relative sizes of the planets and the Sun. Note that most of the planets orbit the Sun within the same plane as that of the Earth's orbit, and they all orbit the Sun in the same direction (anti-clockwise in this Figure which by convention is drawn with the planetary north poles at the top).

(a)

(b)

Can you think of any reasons why the Earth is over three times more dense than Jupiter?

The density of Jupiter is similar to that of the Sun and, like the Sun, it is composed of light elements whereas the Earth comprises a mixture of heavier elements. You may recall that the Earth's core[A] is made primarily of iron, and its surrounding crust[A] and mantle[A] of silicates. To a first approximation, this essential difference in composition is a primary feature that distinguishes *the four inner or terrestrial planets (silicates and iron) from the four outer or major planets (mainly light elements such as hydrogen and helium).*

We now need to remind you about some of the other features of the Solar System and the inner terrestrial planets in particular, gleaned from centuries of observation, greatly supplemented by results from man-made satellites and space probes in recent decades.

Mercury is the smallest planet in the Solar System and is the nearest planet to the Sun. It is smaller even than Jupiter's two largest ice and rock moons (Ganymede and Callisto) and only a little larger than the Earth's Moon. Yet it has a very high density ($5.42 \times 10^3 \, kg \, m^{-3}$); compare it with Mars which is larger than Mercury but has a smaller density. Like all other inner planets Mercury's surface is heavily cratered and is dominated by lava flows. The planet spins on its axis rather slowly: its day is two-thirds of its year.

The next planet in order of increasing distance from the Sun is Venus, which has both mass and density a little less than those of the Earth. Like Mercury, Venus has a volcanic surface; it has high surface temperatures (c. 450–550°C) and a dense atmosphere, mainly of carbon dioxide. The CO_2-rich atmosphere and high surface temperatures may be related since CO_2 is transparent to incoming solar radiation but absorbs long-wavelength outgoing radiation leading to the so-called *greenhouse effect*[A]. Venus spins slowly on its axis in the opposite direction to all the other planets. This is known as *retrograde motion* and is indicated by the minus sign against the axial spin entry for Venus in Table 1. All the planets including Venus orbit the Sun in the same direction (anti-clockwise in Figure 1), and all except Venus spin on their axes in the same direction as their orbits (anti-clockwise in Figure 1).

retrograde motion

The Earth has the highest planetary density but it is also the largest of the inner planets. Its surface has been subjected to geological processes during its lifetime and these have obscured the effects of cratering due to impacts, evidence of which is seen on the surfaces of bodies such as Mercury and the Moon. The Earth has an unusual atmosphere of oxygen and nitrogen which is partly the result of photosynthesis in plants (liberating oxygen) and partly the result of the condensation of water and the removal of soluble gases from the atmosphere into the ocean. *Atmospheric evolution*[A] seems to be more advanced for the Earth than for any other planet. The Earth's large Moon is known to lack present-day 'geological' activity and has a heavily cratered surface that has not been subjected to geological activity for a very long time. Its density is much less than those of Mercury, Venus and Earth. The axial spin period of the Moon is equal to its orbital period around the Earth because the same face of the Moon is 'locked' by gravitational forces towards the Earth. It is the more massive hemisphere of the Moon that faces the Earth. This is the most stable situation gravitationally as you can prove by the simple analogy of letting a bicycle wheel spin freely—it will come to rest with the valve nearest to the centre of the Earth.

The planet Mars is both smaller and less dense than the Earth. It is cratered but shows signs of surface activity: channels, canyons and some of the largest volcanoes yet found in the Solar System. It has a thin atmosphere of water and carbon dioxide, much of which is frozen into polar ice caps. Mars has a day length similar to that of the Earth, though its solar orbital period—its year—is longer.

Before we consider the outer planets, we should see what evidence we have for major variations of *chemical composition* among the inner planets. You already know that density varies with composition because the higher the ratio of heavy to light elements, the greater will be the density.

But will density vary with size between two planets of otherwise identical composition?

You might think intuitively that density should not vary with size. But in fact density varies according to how compressed the material becomes towards the centre of each planet: the more *compressed* the material, the higher the density, other things being equal. This can be described simply, using the fact that the *pressure* (the force acting on a given area) must increase with depth in any planet, because of the weight of overlying material, as follows:

$$P = \rho g d \ \mathrm{N\,m}^{-2}$$

pressure

where P is pressure, g is the acceleration due to gravity and d is the depth of the overlying material with an average density ρ. It is generally true that the pressure at the centre of a large planet (large d) will be greater than at the centre of a small planet *of the same overall composition*. Therefore, if the inner planets all have the same composition, their *average densities* should still vary in proportion to their sizes.

ITQ 1   Look at the radius and density columns in Table 1. How well do you think these data bear out the assumption that the *inner planets* (and the Moon) all have the same composition?

(*ITQ answers begin on p. 92.*)

Note   It is important that you read the answers to all the ITQs in this Block, as some of the answers develop arguments that start in the Main Text.

The four major (outer) planets are very different from the inner planets: they have lower densities and are much larger in size. They are all thought to consist mainly of hydrogen and helium, which are possibly in the metallic state in their high-pressure interiors, giving way outwards to vast gaseous atmospheres. Jupiter and Saturn are both extremely large compared with the Earth but have shorter axial rotation periods which cause very pronounced flattening at the poles and bulging at the equators (Jupiter's polar radius is 94 per cent of its equatorial radius; the Earth's polar radius is 99.7 per cent of its equatorial radius). This general increase in the axial spin rate of planets with distance from the Sun is one of the most puzzling physical features of the Solar System and has defied explanation by theorists of planetary origin. The major planets are well endowed with moons, some larger even than Mercury, and these vary considerably in density from dense iron–silicate materials to light, water-rich 'dirty snowballs'. The four 'Galilean' satellites (the largest moons) of Jupiter are proving to be geologically fascinating. Io, for example, is a little larger and more dense than the Earth's Moon. It has active volcanoes that generate sulphur-rich lava flows and its activity is attributed to heating through the enormous gravitational forces between Io and Jupiter, which raise tides in the solid material of the moon. The other large moons of Jupiter are more distant from the main planet and appear to be frozen and inactive.

Saturn's ring system (*CB*, Plate 2) is nearly 3×10^5 km wide and yet the rings have turned out to be only a few km in thickness. Jupiter and Uranus also have rings, although they are less conspicuous. The rings of all three planets are composed of ice and frozen gases and are thought to be uncondensed moons. All the outer or major planets probably have dense rocky cores; the higher densities of the smaller major planets—Uranus and Neptune—compared with Jupiter and Saturn may reflect the greater importance of dense cores in these planets.

Pluto is the remaining planet not yet mentioned. It is small but of lower density than the terrestrial planets and so would seem to be exceptional. But it is not unlike some of the moons of the major planets and is thought either to have been captured from outside the Solar System or, more probably, to be an escaped moon of the major planet, Neptune.

This description of the Solar System would not be complete without mention of two groups of small, but very important bodies: *comets* and *asteroids*. Comets are low-density objects *some* of which cross the Solar System in a spectacular way, with brilliant shining heads of incandescent gas and long vapour trails, following orbits that cross those of all the planets (Figure 1). These long-period comets pass close to the Sun once in 10^3–10^7 years and are derived from the Oort cloud. This is a region of fully-formed comets (1–50 km diameter), situated outside the orbit of Pluto, about 40 000 AU from the Sun. Comets were observed as early as *c.* 3 000 B.C. by the Chinese, who regarded them with terror, but they are now a harmless source of astronomical fascination. Perhaps the best known short-period comet is Halley's comet, which was last seen in 1910 and is due to be close to the Earth again in 1986. Short-period comets have orbits around the Sun entirely within the known planetary system.

comets   asteroids

Asteroids are another group of bodies ranging from small planets to minute fragments many of which are in a Sun-centred orbit between Mars and Jupiter (Figure 1). In Table 1 you will see that there is a gap at this point in the otherwise regular progression of planetary distances from the Sun. Ceres, of radius 380 km, is the largest known asteroid, but almost 2 000 others have been observed and there are probably millions of tiny fragments. It is almost certain that there were originally several mini-planets in this part of the Solar System that failed to accrete into a single planetary body. Some of these have collided and broken into fragments whose orbits around the Sun were disturbed as a result of these random collisions in such a way that, like the comet shown in Figure 1, their orbits are no longer centred on the Sun. In fact, detailed observations of the orbits of some large and rather exceptional asteroids have shown that they cross the orbital paths of the Earth and other inner planets.

Many exotic blocks or fragments have been observed to *fall* to the Earth on tracks that can be plotted back to the asteroid belt. Others have been *found*—both falls and finds have been collected and studied. Can you remember what we call these exotic objects?

We are referring to an immensely varied group of bodies known as *meteorites*, many of which are thought to have originated in the asteroid belt. These are very important in our studies of the Earth's structure, composition and origin because:

(i)   most of them have radiometric ages greater than those measured from Earth rocks;

(ii)   some of them have compositions that are similar to the inferred total compositions of terrestrial planets;

(iii)   most of them were once inside planets and some may even represent the cores and mantles of layered parent planets;

(iv)   some seem to record the very earliest events that took place in the Solar System, and so give us some clues as to its pre-planetary history.

These points are taken up in the next Sections and again in Block 7 but to complete this Section, we shall take this opportunity to remind you of the main groups of meteorites (see also *CB*, Plate 3).

1   *Stony meteorites* make up about 90 per cent of all known meteorite falls and are sub-divided into chondrites and achondrites.   **stony meteorites**

(a)   *Chondrites* are named after the presence of small silicate, once molten globules or *chondrules* that characterize this group. These meteorites are composed of iron–magnesium silicates, dispersed grains of metallic iron–nickel alloy and the iron sulphide mineral, *troilite* (FeS). Chondrite meteorites are by far the most abundant group; they alone make up about 85 per cent of all known meteorites (*CB*, Plate 3c).   **chondrites** **chondrules** **troilite**

(b)   *Achondrites*, as their name implies, lack chondrules. They also lack abundant metallic grains and are composed of silicate minerals, many of them rich in iron and magnesium. About 8 per cent of known meteorites fall into this category.   **achondrites**

2   *Iron meteorites* are the most spectacular group because they are almost entirely metallic, predominantly composed of an iron–nickel alloy containing between 4 and 20 per cent nickel; they also contain some troilite. They are the most abundant meteorite 'finds' principally because they contrast with all terrestrial rocks; but they are believed to form only a few per cent of total falls (*CB*, Plate 3a).   **iron meteorites**

3   *Stony–iron meteorites* are a mixture in that they consist of a matrix of iron–nickel alloy that surrounds grains or fragments of silicate minerals (see *CB*, Plate 3b). They form about 2 per cent of total falls.   **stony–iron meteorites**

In Section 2.3 we shall be looking again at meteorites and their geochemistry in order to find out what they can tell us about the distribution of elements in the Earth and the Solar System. But first we shall examine some fundamental questions of geochemistry: *where and how were the chemical elements formed and what are their relative abundances*?

One important aim of this Course is to discuss the bulk composition of the Earth and then to show how the elements have become segregated into layers inside our planet. We should like to persuade you that the evidence from the rest of the Solar System is valuable in this context because the Sun and the planets are probably linked by a common origin. In fact, you may remember the *nebular theory*[A] for the origin of the Solar System which was postulated by the Marquis de Laplace in 1796 and which has been developed in various guises during the present century. The idea is that a flat, disc-shaped, rotating cloud of gas and dust (a nebula) gradually contracted and condensed to form the Sun. But in so doing, the disc could not lose its overall *momentum*[A] and therefore rotated progressively more rapidly as it contracted. This led to a ring (or rings) of material being left in stable orbit around the centre (much like Saturn's rings today, but on a larger scale—*CB*, Plate 2) and these condensed into planets. According to this theory and others like it the components of the Solar System all have a common origin and so you might wonder why the Earth does not consist primarily of hydrogen and helium like the Sun. Quite clearly, the Earth is composed of heavier elements (Table 1) and, if we tell you that the Sun also contains some heavy elements (about 2 per cent of its mass), then it is logical to ask if there is any relationship between the composition of the Earth and the relative abundances of these elements in the Sun.

2.2   Cosmic abundances of the elements

Before we begin to set up a model for the Earth's composition in Section 2.3, we shall deal first with the abundances of the elements in the Sun, a typical star. We shall then consider the *origin* of the chemical elements that became incorporated into the Sun and planets when they formed. In fact, the evolution of the entire Universe (or *cosmos*) depends on the same processes that created the chemical elements in the Solar System and this is why we have entitled this Section the *cosmic abundances of the elements*. The point is that most of the matter in the Universe is concentrated in stars, and what we need is a method of chemically analysing stars. The most convenient star to demonstrate this technique on is our own—the Sun.

cosmic element abundances

2.2.1   Solar spectroscopy

The only way in which we can carry out any measurements of the Sun's composition is by studying its radiation by using the techniques of *solar spectroscopy*[A]. In fact, the measurements we make are of the Sun's *atmospheric* composition: Figure 2 shows this atmosphere, which extends thousands of kilometres from the visible surface of the Sun. What happens is this: the immensely hot interior of the Sun (the visible disc in Figure 2) emits a *continuous spectrum*[A] of radiation; that is, one in which all frequencies are represented. But this radiation has to pass out through the lower atmosphere, which itself is very hot (*c*. 6 000 °C), but which is much cooler than the interior of the Sun (several million degrees). As a result, the atoms of elements in the lower atmosphere *absorb* some of this radiation at frequencies appropriate to their own ionization energies and that gives the thousands of dark lines which can be found in the spectrum of the Sun. These are the *Fraunhofer lines*[A], first observed in 1814 and named after their discoverer, J. Fraunhofer.

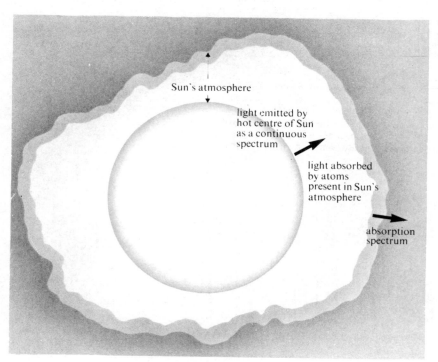

FIGURE 2   The Sun's absorption spectrum is generated in the solar atmosphere where light emitted from the visible disc is absorbed in spectral bands corresponding to the radiation frequencies of the elements present.

At this point it is important to make sure that you realize the difference between *emission* and *absorption spectra*. If you were to heat any element on the Earth to the temperature of the Sun's lower atmosphere, you would obtain a characteristic *emission* line-spectrum for it. This is because, during heating, loosely held electrons continuously move to higher energy levels and then fall back to their original levels, re-emitting the extra thermal energy as electromagnetic radiation of a specific and characteristic frequency (f) defined by the photon energy (E) relationship:

$$E = hf$$

where h is *Planck's constant*[A].

So why do we obtain *absorption* lines on the Sun's spectrum?

This is because the Sun's lower atmosphere is so much cooler than the interior that the atoms *absorb* radiation passing through it from the solar surface. They are able to do this by their electrons jumping to higher energy levels in the same way as when heated on Earth. Of course, as in the production of an emission spectrum, electrons must then fall back to lower energy levels, re-emitting radiation. So, you may still wonder, why are there absorption lines, because for every photon of light absorbed one should be re-emitted? The answer is that the number of absorption electron jumps (to higher levels) exceeds the number of emission jumps (to lower levels) because the atoms become much hotter as they collide with each other in the atmosphere. Studies of the solar corona during eclipses have shown that the upper atmosphere may reach one million degrees centigrade, much hotter than the Sun's yellow visible surface and lower atmosphere, and these high temperatures are the result of frictional heating during collision between the rapidly moving atoms. So gases in the Sun's atmosphere become progressively more 'excited', that is, the efficiency of electron jumps to higher energy levels increases with temperature and, eventually some electrons are lost entirely. The material is now *ionized*: the net result is energy *absorption* and the ionized gases stream away from the Sun to be replaced by cooler material from the Sun's surface.

You should now understand how the Fraunhofer lines are formed. You should also appreciate that the frequencies (f) of the observed lines characterize the chemical elements present in the atmosphere. These elements cause the selective absorption lines of the continuous spectrum which itself originates in the Sun's interior. *CB*, Plate 4 illustrates the solar absorption spectrum for the frequency range of visible light, and the more prominent lines are identified by letters in Figure 3. In 1882, H. A. Rowland photographed and published a 40-foot long map of the Sun's spectrum. Two small parts of it appear in Figure 4, with the dark lines matched to sodium and calcium labelled with the same letters as in Figure 3: so you can appreciate that this map is a very detailed record of the Sun's spectrum.

FIGURE 3   Diagram of the solar absorption spectrum with the most prominent lines labelled (as by Fraunhofer) as follows: A, B = oxygen; C, F = hydrogen; D = sodium; E, G = iron; H, K = calcium (1 nm = 10^{-9} m). Note that wavelength decreases from left to right in CB, Plate 4, opposite to that shown here.

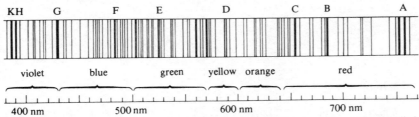

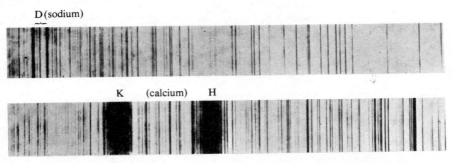

FIGURE 4   Two small parts of H. A. Rowland's original large-scale map of the solar spectrum (D = sodium; H, K = calcium).

How can we use the Fraunhofer lines in the solar spectrum to determine

(a)   the presence of different elements in the Sun's atmosphere?

(b)   their relative abundances?

Your understanding of absorption and emission spectra should have enabled you to answer (a), remembering that, for a particular level of excitation (that is, at comparable temperatures), an element will *emit* or *absorb* the *same* frequencies, depending on circumstances. So, if you compare the known *emission* line-spectrum of a particular element with the *absorption* line-spectrum of the Sun, then the bright lines on the emission spectrum should match up with some of the dark Fraunhofer lines on the solar spectrum if the element is present.

In practice, the absorption spectrum of the Sun and the emission spectrum of a laboratory source of one element (usually iron) are recorded on the same photographic plate, as in Figure 5.

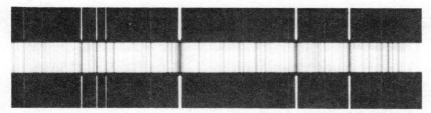

FIGURE 5   A portion of the solar absorption spectrum with an emission spectrum of iron (top and bottom) for comparison.

Take a close look at Figure 5 to see whether the emission lines (light) coincide with any of the dark absorption lines.

Yes, obviously they do, and so we know immediately that iron must be present in the solar atmosphere. (This is a convenient moment to remind you that there are many lines in the emission spectra of individual atoms because there are many electron energy levels between which electrons may jump.)

In theory, this process could be carried out for all the other elements present in the solar spectrum, but it would be very time-consuming and so the other frequencies present are found by *calibrating* the spectrum with the accurately known frequency values of the iron emission lines (Figure 5). As the frequencies of emission (and hence absorption) lines are known for all elements, the presence or absence of their spectral lines tells us whether or not they occur in the Sun's atmosphere. In this way, the majority of the known chemical elements have been identified in the solar atmosphere.

Now for part (b) of the question above: how do we make quantitative estimates of element abundances?

In brief, the *intensity* of a spectral absorption or emission line can be determined using a series of photographic plates exposed for different lengths of time. The intensity of the line will depend partly upon how much of the element is present and partly upon the temperature. It is possible to separate these effects to produce a set of corrected absorption lines whose intensities yield the relative abundances of the elements present. This technique is known as *spectrochemical analysis.*

spectrochemical analysis

This basic procedure has been used to determine element abundances not only in the Sun but in many other stars as well. On this basis, various estimates of the cosmic abundances of the elements have been made, which take into account the fact that the atmospheres of most stars have similar element abundances. A particular compilation by A. G. W. Cameron appears in Figure 6. Of course, it is the *relative abundances* of elements that are plotted; it would be possible, however, to determine the *absolute* quantity of all the elements in a star given the absolute amount of any single element. Since we do not know the absolute abundance of any element, for convenience, the relative cosmic abundances of the elements are usually quoted by setting the abundance of the common 'heavy' element, silicon, at 10^6 atoms. The abundances of all the other elements are then expressed also as numbers of atoms, based on the relative intensity (corrected for temperature) of their absorption lines compared with that of silicon.

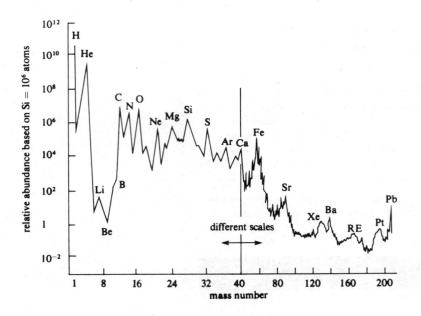

FIGURE 6 The cosmic abundances of the elements expressed relative to the abundance of silicon ($Si = 10^6$ atoms). Note that the vertical scale is logarithmic. (Further explanation is given in the text.) RE = rare earth elements.

The main features of Figure 6 are:

(i)   In terms of numbers of atoms, H and He are more abundant than any other elements in most stellar spectra by several orders of magnitude. For example, the Sun's atmosphere probably contains 70 per cent *by mass* of hydrogen, which has the least dense atoms of all the elements, and about 28 per cent by mass of helium. Just how representative these proportions might be for the Sun as a whole will be discussed shortly.

(ii)   Although the abundances of the elements decrease as mass number increases, this decrease is not smooth. There are relatively high abundances for elements with even numbers of protons (even *atomic numbers*[A]) and atomic masses (*mass numbers*[A]) that are multiples of 4—for example, $^{12}_{6}C$, $^{16}_{8}O$, $^{20}_{10}Ne$, $^{24}_{12}Mg$, $^{28}_{14}Si$, $^{32}_{16}S$ and $^{56}_{26}Fe$.

So the remaining 2 per cent of the Sun's atmospheric mass (98 per cent being H and He) is dominated by eight elements, the seven above and nitrogen (Figure 6), but also contains other elements in smaller amounts. This is extremely interesting because, as you may have realized already, five of these elements, O, Mg, Si, S and Fe, are probably the most abundant in the Earth.

We will come back to this a little later, but first it is useful to give you an insight into why these elements are relatively abundant and, indeed, what factors control the so-called cosmic abundances of the elements. For the moment we are assuming that the chemical compositions of stellar atmospheres reflect the compositions of their interiors. This is true, but only to a first approximation, as you will see at the end of Section 2.2.2.

2.2.2   Origin of the chemical elements: nucleosynthesis

To understand the formation and relative abundances of the chemical elements in stars, we must know something about the kinds of *nucleosynthesis* reactions (reactions that build up progressively more massive nuclei) that continuously produce their heat and light energy.

nucleosynthesis

Can you remember how the Sun 'manufactures' this energy?

We are referring here to *nuclear fusion*[A] reactions that take place at temperatures of millions of degrees centigrade inside stars, reactions that combine nuclei of small atomic mass into nuclei of larger mass, at the same time releasing energy. For example, the following reaction, converting hydrogen into helium, is arguably the most important inside all stars:

$$^{2}_{1}H + ^{2}_{1}H \longrightarrow ^{3}_{2}He + ^{1}_{0}n + 3.4 \, MeV \text{ of energy} \qquad (1)$$

This reaction combines two heavy hydrogen isotopes to produce one light helium isotope and a neutron. $^{2}_{1}H$ is known as deuterium and its natural abundance, as seen on the Earth, is only 0.015 per cent that of normal hydrogen, $^{1}_{1}H$.

In equation 1, the total mass of the two reactant atoms very slightly exceeds that of the two products. The small amount of mass which is lost during the reaction appears in the form of *energy* (from the Einstein relation: $E = mc^2$ where c is the speed of light and E is the energy output resulting from a loss in mass, m). In physical terms, we say that the constituent protons and neutrons of the helium nucleus are more *tightly bound* than in the hydrogen nucleus and it is the *unbinding energy*[A] which would be needed to drive equation 1 to the left that is released as it goes to the right (see Figure 7, p. 18).

It might help you to understand the conversion of hydrogen into helium inside stars if we tell you that equation 1 is just one step in a *chain* of reactions that builds up *one* ^4_2He nucleus from every *four* ^1_1H nuclei. The nucleus of the hydrogen atom contains just a single proton and so the complete conversion from ^1_1H to ^4_2He is known as a *p–p chain*; where p stands for ^1_1H:

p–p chain

$$p + p \longrightarrow {}^2_1H + e^+ \qquad (2)$$

$$^2_1H + {}^2_1H \longrightarrow {}^3_2He + n \qquad (3)$$

$$^3_2He + {}^3_2He \longrightarrow {}^4_2He + p + p \qquad (4)$$

This is a three-step chain reaction; each step releases energy and other particles such as positrons (e^+). Note that the second step is the same as in equation 1. Energy-producing processes involving combination of hydrogen nuclei are often known as *hydrogen burning*.

hydrogen burning

At higher temperatures, which are unlikely to be achieved inside the Sun, helium nuclei may start to combine in a process known as *helium burning*. Particularly noteworthy is the reaction that produces the most abundant isotope of carbon, $^{12}_6C$:

helium burning

$$^4_2He + {}^4_2He \longrightarrow {}^8_4Be_{(unstable)}; \; {}^8_4Be + {}^4_2He \longrightarrow {}^{12}_6C_{(stable)} \qquad (5)$$

The importance of this reaction inside stars helps to explain the high abundance of carbon in Figure 6, though the carbon inside the Sun must have been incorporated from other stars and was not formed in the Sun.

> Does this give you a clue to the preponderance of other elements whose mass numbers are multiples of four?

As you may have noted, additions of helium nuclei *beyond carbon* will produce most important isotopes of elements that are abundant in Figure 6: $^{16}_8O$, $^{20}_{10}Ne$, $^{24}_{12}Mg$, $^{28}_{14}Si$, $^{32}_{16}S$, and so on, up to $^{56}_{26}Fe$. (To reach the last isotope mentioned here, two protons are also converted into neutrons; remember that it is the atomic number that characterizes an element—in this case, 26 for iron.)

The formation of $^{12}_6C$ in stars larger than the Sun actually triggers another cycle of reactions which, in addition to producing more helium, also generate various isotopes of nitrogen and oxygen. The details need not concern us here, except that the process is known as the *CNO cycle* (carbon–nitrogen–oxygen cycle); oxygen and carbon are slightly more abundant than nitrogen in the products (Figure 6).

CNO cycle

All the other elements *up to mass number 56* can be produced by nuclear reactions inside sufficiently large stars by the addition of various combinations of protons and neutrons to the main participant isotopes of the CNO cycle. But the final abundances of all the isotopes whose mass numbers are *not* multiples of four are always lower than those isotopes which can be produced directly from additions of helium nuclei (see Figure 6). Given suitable temperatures, all the fusion reactions up to mass number 56 are self-sustaining and *exothermic*[A]. But the necessary temperatures increase markedly: whereas the p–p chain reactions occur at about $2 \times 10^8 \, °C$ and the CNO cycle is effective at about $5 \times 10^8 \, °C$, a reaction such as:

$$^{28}_{14}Si + {}^{28}_{14}Si \longrightarrow {}^{56}_{28}Ni_{(unstable)} \longrightarrow {}^{56}_{26}Fe_{(stable)} + e^+ + e^+ \qquad (6)$$

requires temperatures approaching $4 \times 10^9 \, °C$ to be achieved. As temperatures increase, and increasingly heavy nuclei are fused, the reactions become less exothermic; that is, they release progressively less energy. To see why this should be so, take a look at Figure 7, in which the average unbinding energy per nuclear particle (i.e. the constituent protons and neutrons of the nucleus) is plotted against the mass numbers of naturally occurring nuclei. As we climb the left-hand side of the diagram, fusing atoms up to a mass number of 56, the average unbinding energy for each heavier nucleus *increases. The difference between the unbinding energies of the nuclei on either side of equations 1–6 (products minus reactants) determines the*

17

amount of energy released. Clearly, the p–p chain (equations 2–4), which uses $_1^1$H to create $_2^4$He, releases much more energy than the CNO cycle which, in turn, releases more energy than equation 6.

ITQ 2 So far we have seen that the combination of light elements into heavier elements releases energy and this is because the average unbinding energy per nuclear particle increases with mass number up to 56. What do you think is going to happen to the energy balance as more massive nuclei than iron (mass number = 56) are 'fused'? Use Figure 7 to help you in answering this question.

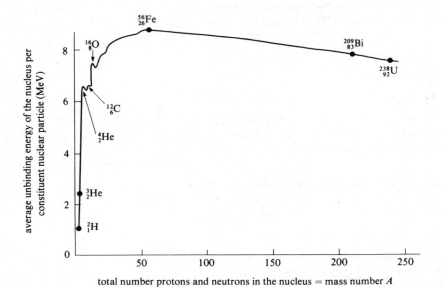

FIGURE 7 The nuclear 'unbinding' energy curve: plot of the energy required to break apart the nuclei of different isotopes (per constituent nuclear particle) against mass number. The higher up the curve a nucleus plots, the more stable it is.

$_{26}^{56}$Fe has the greatest nuclear binding energy per constituent particle of all elements and, although there are many stable isotopes of elements up to bismuth ($_{83}^{209}$Bi), most of the remaining heavy isotopes have large unstable nuclei which break up in two important ways. Some, the least stable isotopes, spontaneously split apart into random mixtures of smaller atomic particles by *nuclear fission*[A]. Others decrease their nuclear mass in a more controlled way by *radioactive decay*[A]; for example, by losing α-particles (an α-particle is equivalent to the helium nucleus $_2^4$He).

Both processes *release* energy (unbinding energy increases from the right towards iron in Figure 7) and nuclear fission releases much more energy than alpha decay– hence the use of fission reactors by electricity generating boards. Some isotopes undergo both fission and α-decay: in a mass of $_{92}^{238}$U atoms, for example, one will undergo fission for every two million that decay. Clearly, given long enough, there would be no atoms of all these heavy elements left unless they are replenished in some way.

To explain the presence of heavy elements inside stars such as the Sun (and in the Earth) we need a process that will supply a massive *input* of energy to the endothermic *fusion* reactions on the right of Figure 7 (look back at the answer to ITQ 2). This is quite contrary to the normal activity of stars, which is to generate energy by exothermic reactions. We shall give you a probable explanation for the production of heavy elements in a moment. But first, it is useful to consider what is generally believed to happen during the lifetime of a star. It will be obvious by now that the chemistry of the Universe is dominated by hydrogen, closely followed by helium. Current cosmological theories explain the superabundance of H and He in the Universe as stemming from a 'big bang' event which occurred about 15–20 Ga ago. Except for traces of a few light elements, the matter resulting from this event was dominated by these two elements. All the stars that condensed from this primordial matter were, therefore, rich in hydrogen, together with helium. In common with the evolution of all stars, they would have started their lives as dispersed 'clouds' of low-pressure gas which gradually contracted due to gravitational forces between the gas molecules. As the gas molecules became more confined, their temperature rose until the conditions at the centre were right for hydrogen to burn (equations 2–4). This released energy and arrested gravitational collapse so long as 'burnable' nuclear fuel was present.

Most of the lifetime of any star (its so-called *main sequence*) is spent in producing helium from hydrogen in a progressively expanding sphere around the star's centre. Not all the hydrogen fuel is burnt, because temperatures in the outer parts of the star never become sufficiently high. But large stars develop higher internal temperatures and get through their supply of fusible hydrogen much more rapidly than small stars: the Sun is not large and will take about 10 Ga to complete its active life cycle (it is now about halfway through this life cycle). Of course, once hydrogen has 'burnt' to helium, the temperature in a large star may become sufficiently high for helium to burn, and then for other elements to burn, and so on up to the production of iron.

The minimum mass for the temperature to be achieved at which elements with nuclei heavier than iron can be produced is about six times the Sun's mass. Whereas the Sun will probably just burn out, a star of about six times its mass may evolve through a 'red giant' stage towards a massive terminal explosion, known as a *supernova*. A critical state is reached in which the star has an iron-rich core at a temperature of about 10^{10} °C. Here nuclei close to the mass of iron, including iron, begin to break down again (by fission) into simpler particles: α-particles, neutrons, etc. Such a star will have been stratified by previous nuclear fusion into an iron core surrounded by shells of other elements. The rapid fission of some core material rapidly *absorbs* a great deal of energy and this is provided by contraction of the whole star. This brings previously unfused material to higher temperatures where fusion can take place. A runaway situation develops in which burning raises the temperature, and this, in turn, increases the rate of burning. The star has no time to adjust and much of the remaining fuel is consumed in seconds, blasting the outer shells away into interstellar space at high velocity. Meanwhile, the rise in temperature has *supplied the energy needed for the heavy isotopes to be synthesized*. These can only be produced in supernovae and at no other times or places in the Universe. In an intense bombardment of protons, neutrons and helium nuclei the existing elements, built up by helium additions inside the star, may be transmuted into many whose nuclear masses are not multiples of four. For example, starting from iron nuclei, elements of higher atomic number up to lead can be produced by the progressive addition of neutrons.

supernova

The core of a large star is thought to collapse very rapidly during a supernova, producing a dense 'neutron star' or 'pulsar' and, in special cases, it may yield one of those famous 'black holes' where gravity is extremely strong. The surrounding products are rapidly disseminated, as incandescent gaseous clouds, throughout interstellar space by the force of the supernova blast. *CB*, Plate 5 shows the Crab Nebula, perhaps the best-known supernova, first observed by the Chinese in 1054, and from which rapidly dispersing fragments are still visible.

Now we can begin to think again about how the elements heavier than hydrogen and helium became incorporated into our star, the Sun. The Universe is between 15 and 20 Ga old but the Sun is only about 5 Ga old. In our Galaxy there are about 10^{11} stars and, perhaps, three supernovae per century. Although 2 per cent of the Sun's mass is not hydrogen and helium, it has probably not advanced beyond p–p chain fusion reactions (equations 2–4).

> So where might these heavy elements have come from, and how do you think they were incorporated into the Sun?

The answer lies in previous supernovae! Remember, as explained above, that the *larger* the star, the *heavier* the elements that it may synthesize. It takes a star at least six times as massive as the Sun to produce iron and other heavier nuclei during a supernova. Such massive stars burn through their supply of fusible elements rapidly because they can more easily develop high internal temperatures. Evidently stars are being born and are dying all the time and the Sun is a relatively young star (5 Ga old) which formed at least 10 Ga after the primordial matter of the Universe first condensed into stars. During that first 10 Ga there may have been as many as 3×10^8 supernovae in our Galaxy (at the rate of 3 per century) *each of which would enrich the galactic reservoir of H and He with heavy elements*. On this basis we can predict that the younger the star, the more heavy elements it is likely to contain— and this is just what is observed. Viewed on a cosmic scale, a few per cent of the primordial H and He have been converted into other elements. So the original contracting cloud (or nebula) of material that formed the Sun *already contained small quantities of C, N, O, Ne, Mg, Si, S, Fe and all the other elements observed in its spectrum.*

ITQ 3 Consider concentrations by mass in the Sun and ignore the loss in mass due to fusion reactions. What changes have there been since the Sun formed in the concentrations of:

(i)   hydrogen?
(ii) helium?
(iii) heavier elements?

Now you should appreciate where the heavy elements in the Sun and, indeed, the rest of the Solar System, *including the Earth*, came from. All the elements that we normally associate with rocks and that probably form the inner planets are products of late-stage reactions in the lives of large stars, which became supernovae, and the relative abundances of these elements must reflect 'local' events in our Galaxy. In other words, they are a distinctive mixture of the products of nearby supernovae which exploded earlier than the formation of our Solar System. You may have realized that all stars must have slightly different compositions depending on when and where they formed. Therefore, there really is no one 'cosmic abundance of the elements' because so many different nucleosynthesis processes contribute elements in different proportions and in different parts of the Universe. Figure 6 displays the average of many stellar spectral analyses.

> For simplicity, imagine that one particular supernova event gave rise to most of the heavy elements in our Solar System. Can you think of any way in which we might find out *when* that event took place?

Well, that's quite a difficult question, but you may have thought that radioactive parent and daughter isotopes provide some clues. Terrestrial *rocks* can be *dated*[A] by analysing them chemically to find the amounts of parent and daughter, whose ratio is time dependent. One particular example is the decay of $^{235}_{92}U$ to $^{207}_{82}Pb$, in which half the parent atoms decay every 704 Ma (the half-life). The Earth itself is dated at *c.* 4 600 Ma by measuring the amounts of these two isotopes now present and then correcting for the estimated amount of $^{207}_{82}Pb$ which was originally incorporated into the Earth when it formed. In other words, for this isotope:

$$Pb_{now} = Pb_{radiogenic} + Pb_{original}$$

and $Pb_{radiogenic}$ is used to find the age of the Earth.

So that allows us to answer our earlier question about the timing of a pre-Solar System supernova: if we could find the amount of $^{207}_{82}Pb$, for example, produced during the appropriate nucleosynthesis processes (for example, from observation of uranium-free meteorites), then the supernova event could be dated. In fact, this proves to be difficult to carry out for elements that decay slowly, such as isotopes of uranium. What we need is an isotope that was decaying more rapidly during the critical time interval between our supernova and the formation of the Earth. During the 1970s, evidence that *short-lived radioactive isotopes* were incorporated into the Solar System came to light: most notably, the isotope $^{26}_{13}Al$ which decays to $^{26}_{12}Mg$ with a half-life of 7.2×10^5 years. The evidence for this comes from the excess $^{26}_{12}Mg$ that has been found in Moon rocks and meteorites.

> How much of the $^{26}_{13}Al$ originally present in the Solar System should we find today?

A simple calculation shows that so many half-lives have elapsed (*c.* 6 000) that none will be left. Indeed, for significant amounts to have been incorporated, the Solar System must have formed within a few million years after a supernova event manufactured this isotope of aluminium. Detailed studies of the daughter products of short-lived radioactive isotopes, such as $^{26}_{13}Al$, have yielded a most spectacular conclusion: that *supernovae exploded in our part of the Galaxy about 100 Ma and again only about 1 Ma before the planets in the Solar System were formed*. So the elements from which the Earth is made were 'manufactured' not in one, but in at least two and probably many more earlier supernovae which took place before 4 600 Ma ago. The elements were incorporated into the Sun and planets by various processes of accretion, which are introduced in Section 2.3 and are described more fully in Block 7.

Let us now return to the assumption (made at the end of Section 2.1) that the bulk chemical composition of the Earth and the 2 per cent of heavy elements in the Sun are similar. This stemmed from *the proposition that the Earth and the Sun had a common origin in the primitive solar nebula*. We have shown that the assumption is correct in a qualitative sense for the five most abundant elements in the Earth: O,

Mg, Si, S and Fe. The Earth and the other inner planets contain very little hydrogen, helium, carbon, nitrogen and neon, compared with their abundances in stellar spectra (Figure 6)—despite all the carbon and nitrogen in the Earth's atmosphere, oceans and living organisms. Section 2.3 examines this difference and suggests that these light elements were either lost as gases or were not incorporated into the Earth. Section 2.3 also considers the geochemistry of meteorites, which are the only other available simple, well-studied materials from within the Solar System.

2.3 Compositions of the Earth, meteorites and the primitive solar nebula

Figure 8 summarizes the best available estimates that we have for the Earth's internal structure and composition.

> Can you recall, from your Science Foundation Course, the most important source of evidence we have about the internal structure and composition of the Earth?

We are referring here to *seismology*[A], the study of earthquake waves and the way they pass through the Earth. Earthquake wave velocities are a function of various physical properties of the media through which they pass, notably density. The model for the Earth's composition shown in Figure 8 is based partly on seismological data. The way in which the physical properties of the deep Earth are determined is described more fully in Block 2, in which we shall be looking at the details of Figure 8 again.

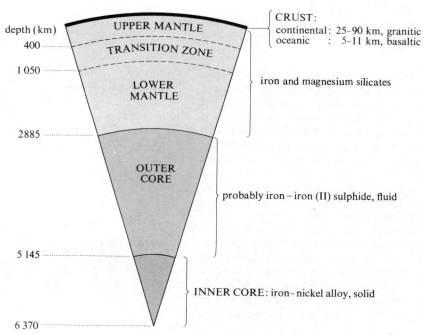

FIGURE 8 Summary of the Earth's layers and their probable (generalized) compositions. Note that these generalizations will be discussed and refined in Block 2, Section 4.

But seismological data provide only the crudest estimates of chemical composition. In order to improve our knowledge of the Earth's composition, especially the composition of the deep, inaccessible layers, we also need good estimates of the abundances of heavy elements in the *primitive solar nebula* (*PSN*). Figure 6 provides one possible estimate, but to use it quantitatively relies on a knowledge of the absolute abundance of a key element, such as silicon. Remember also that Figure 6 is based on spectral observations of many stars, not just the Sun. A better constraint on the composition of the Earth's layers can be obtained from *meteorites*, first because we can analyse them directly and quantitatively; second because, like the Earth, they concentrated heavy elements from the PSN where they formed. At the end of Section 2.1 we revised, very briefly, the main types of meteorite: irons, achondrites, stony-irons and chondrites. These meteorites are made of silicate minerals, metallic iron–nickel alloy and iron sulphide in various proportions.

primitive solar nebula (PSN)

> Now look again at Figure 8. Can you suggest what sorts of meteorites might provide the best guides to the composition of different layers in the Earth?

Starting with the Earth's *core*, it is quite clear that *iron meteorites* are a good guide to composition, although seismological and other data indicate that the Earth's core may contain more sulphur than most iron meteorites. This sulphur is probably concentrated into the liquid outer core (you will learn more about this in Block 2). *Achondrite* meteorites are used to model the Earth's mantle and oceanic crust because they contain mainly iron and magnesium silicate minerals. A few are strictly *peridotitic*[A] (this will be discussed in Section 4 and Block 2), but most contain more calcium and aluminium, which makes them *basaltic*[A], like the Earth's ocean crust. There are no known meteorite equivalents of the Earth's continental crust and there are no zones within the Earth that seem to have the mixed metal–silicate composition of chondrites (silicate with dispersed metal) or stony–iron meteorites (metal matrix with silicate inclusions, *CB*, Plate 3b).

> If we are right to use iron meteorites and achondrites to model the Earth's core, mantle and oceanic crust, what relationship might these meteorites have had to each other in their parent mini-planets?

You might have been encouraged to *hypothesize* that iron meteorites may represent the cores, and the different achondrite meteorites the mantles and basaltic crusts of planetary bodies. Earlier, we noted that quite large mini-planets (up to 380 km radius) have been observed in the probable source region of meteorites (the asteroid belt) and a part of our hypothesis must be that some of these bodies have collided and broken up to yield iron and achondrite meteorites. Stony–iron meteorites have a place in this hypothesis, too, because they contain a lot of metal and are usually interpreted as deriving from an iron–silicate boundary region. Clearly, there is good reason to suppose that the parent meteorite mini-planets had layered structures comprising separate iron and silicate regions, like the Earth.

> There are various ways in which layered structures might be developed in planets. Can you think of at least one?

This problem of *planetary layering*[A] was discussed in the Science Foundation Course. You may recall that the planets could have been relatively *cold* and *homogeneous* to start with but that they were heated to the point where melting took place inside. At this point, materials might separate according to their density, and this is one possibility for the formation of dense iron-rich cores such as that of the Earth. This mode of planetary accretion implies that a powerful heat source is needed to cause melting (1 000–1 500 °C) and the development of layers. Possible heat sources are gravitational and radioactive processes. An important example of *gravitational heating* occurs during accretion when impacting particles lose their kinetic energy which is converted partly into heat. All other things being equal, the magnitude of the heating effect will be proportional to the strength of the gravitational field, which grows with time. Therefore gravitational heating is more important in massive bodies. *Radioactive heating* depends on the concentration of heat-producing radio-isotopes in the planet. But it also related to planetary size because large bodies have a smaller surface area in relation to volume than do small bodies (for a sphere, the ratio of surface area to volume is proportional to 1/radius). Therefore, less of the heat produced inside a large body can be radiated from the surface in a given time interval. So both gravitational and radioactive heating are important in large planetary bodies.

homogeneous accretion

gravitational heating

radioactive heating

Whereas these heat sources can account for melting and hence for the layering of the inner planets, it has been shown that bodies of about the size of the largest asteroids would have been unlikely to reach melting conditions unless significant amounts of short-lived radioactive isotopes such as $^{26}_{13}$Al (see Section 2.2.2) had been incorporated. If those isotopes were present, short-lived radioactivities would have provided a highly effective means of heating the early planets.

The obvious alternative to homogeneous accretion is that the iron-rich cores grew first and only later became surrounded by a layer of silicate material. In other words, the planets were the products of *heterogeneous accretion*. This must mean that the PSN was so hot that all the elements were in the *gaseous* state and that, as its temperature fell, iron condensed (i.e. became solid) before the silicates*. Hot, heterogeneous planetary accretion provides a means of producing layered structures directly from the PSN and so removes the immediate need for internal

heterogeneous accretion

---

* Only about 1 800 °C is needed for all elements—even iron—to be gaseous at the very low pressures that must be typical of a nebula.

heat sources. However, it is unlikely that the Earth's original heat was sufficient to maintain its dynamic, convecting mantle, its fluid outer core and its observed heat flow throughout its entire history (this will be discussed in Block 4), and so gravitational and/or radioactive heating are likely to have been important in the Earth's evolution.

We have identified achondrites, irons and stony–iron meteorites that apparently came from layered planetary bodies like the Earth. This leaves the *chondrites*, by far the most abundant group, consisting mainly of silicates, with some metal. As we shall show you in a moment, the chemical compositions of chondrites are such that, if they were heated and separated into iron-rich and silicate components, the proportions of these components would be very approximately the same as in the Earth's core and mantle (as deduced seismically). You might be wondering how the chondrites managed to escape whatever process caused all the other planets and meteorite mini-planets to become layered. Well, we cannot answer that question with any certainty, but it might help you to consider the following possibilities.

All meteorites, including chondrites, have isotopic ages close to 4 600 Ma. This means that they were *either* last heated then *or* that they condensed at that time and were not subject to long-term heating after accretion. Given that all the meteorite parent mini-planets were small, this would be consistent with either of the accretion processes. If the planets were produced by cold, homogeneous accretion, then the chondrites would represent the smallest mini-planets, which could have escaped heating and, therefore, layering. (Alternatively, they could be parts of unheated, near-surface layers of larger planets.) If the planets were accreted heterogeneously, then the chondrites must have formed from primary material that was not accreted into planets. This is because they contain both metal and silicate which must have condensed together in a part of the PSN that was cooler than the part where the inner planets and other meteorites formed. Whether accretion was heterogeneous or homogeneous, it seems likely that chondrites have a primitive nature and probably retain a unique record of the heavy element composition of the PSN.

We can now say that, of all groups of meteorites, *chondrites provide the best chance of estimating the relative abundances of heavy elements in the PSN.* The next step is to consider the geochemistry of chondrites in more detail, to compare their composition with that of heavy elements in the cosmos (Figure 6) and the accessible parts of the Earth.

2.3.1 Chondritic meteorites

We have been talking about chondrites as a rather uniform group of meteorites but, in detail, they are quite varied in both their mineralogical and chemical compositions. First, we shall look at their *chemical compositions.* Chondrites fall into five groups on the basis of ratios between four of their most abundant chemical constituents: iron, silicon, oxygen and sulphur (see Figure 9). These groups are thought to reflect sources with different states of *oxidation*[A] and iron content.

> Given that iron is found in meteorites in three forms: as the metal, as iron sulphide (FeS) and as iron silicate (e.g. Fe_2SiO_4), which of these forms is the most oxidized?

Iron occurs in the relatively oxidized Fe(II) state in Fe_2SiO_4 and FeS, and in the relatively reduced state in iron metal. But for the purposes of classifying meteorites we group together iron sulphide and metallic iron (plotted up the y-axis in Figure 9) and plot iron, *combined with oxygen,* as silicate, along the x-axis.

> **ITQ 4** To make sure you understand Figure 9, can you arrange the five chondrite meteorite groups in order:
>
> (i) according to their relative content of iron in silicate compared with that in sulphide and as metal?
>
> (ii) according to their total iron content?

So chondrite meteorites are amenable to chemical classification and their compositions vary down Table 2, where the H-, L- and LL-groups* have been

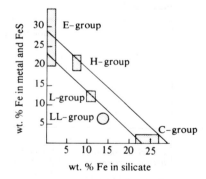

FIGURE 9 Relationship between reduced (metallic and sulphide) iron and oxidized (silicate) iron in the different chondrite groups. Analyses lying on the same diagonal line have the same total iron content. (The E- and C-groups are defined in Table 2; the H-, L- and LL-groups are ordinary chondrites.)

---

* Following ITQ 4 it should not surprise you to learn that the letters H, L and LL stand for 'high-iron', 'low-iron' and 'low-iron, low-metal' groups (see also Figure 9).

One independent factor that increases our confidence that the other inner planets are broadly chondritic in composition is the relative abundances of elements resulting from nucleosynthesis and now seen in the cosmic element abundance curve (Figure 6). This indicates that, after the four abundant light elements, hydrogen, helium, carbon and nitrogen, and the unreactive gas neon, the principal heavier elements present to form the inner planets and meteorites should be O, Mg, Si, S and Fe, and these are the five abundant elements in chondrites. The elements H, He, C, N and Ne, not present in large concentrations in the inner planets and meteorites, must *either* have been lost as gases (carbon as CO or CO_2) during accretion *or* not have been condensed around these bodies in cosmic proportions. Oxygen was 'retained' more effectively than carbon, nitrogen and neon, probably because much was combined into silicates.

2.4 Summary of Section 2

(a)   The principal members of the Solar System are the Sun, four small, dense inner (or minor) planets, four large, low density outer (or major) planets, and Pluto.

(b)   Among the minor planets, density increases in proportion to size, suggesting similar (chondritic) planetary compositions, except for Mercury, which is anomalously dense and, therefore, likely to be iron-rich (see the answer to ITQ 1).

(c)   The so-called 'cosmic abundances of the elements' may be determined by analysing the absorption spectra of stellar atmospheres, which closely reflect the composition of stellar interiors. It is found that, for the Sun, about 98 per cent (by mass) is hydrogen and helium and that the remaining 2 per cent of heavy elements show a characteristic pattern of decreasing abundance with increasing mass number. However, elements with mass numbers that are multiples of 4 are more abundant than the rest.

(d)   The relative abundances of the elements in stars are accounted for by simple processes of nuclear fusion. These release energy in exothermic fusion reactions that create elements up to $^{56}_{26}Fe$. They build up nuclei in stages, first from hydrogen to helium (the p–p chain), then, in large stars, from helium to carbon, and progressively by further additions of helium nuclei, produce abundant elements with mass numbers that are multiples of 4.

(e)   Late in the life of large stars, internal temperatures are sufficiently high for more complex endothermic reactions to synthesize rapidly all the naturally-occurring elements heavier than iron. Many of them are unstable and are added to the interstellar gas and dust reservoir through supernovae expositions. The Sun (condensed from such a reservoir) is relatively small and its present-day element abundances are likely to be unchanged since formation about 5 Ga ago from the primitive solar nebula (PSN) except for a relative increase in He at the expense of H.

(f)   Planetary materials contain daughter products from extinct short-lived radioactive isotopes, such as $^{26}_{13}Al$, that have not been manufactured in the Sun. This suggests that nearby supernovae must have immediately preceded the contraction of the PSN that formed the Sun and planets.

(g)   Meteorites, which are thought to derive from the asteroid belt between the orbits of Mars and Jupiter, can be divided into:

(i)   iron and achondrite groups, which may have been the cores and mantles of parent mini-planets that were layered like the Earth, and

(ii)   a chondritic group with variable abundances of relatively volatile elements (such as alkalis) that reach a maximum in the carbonaceous chondrite sub-group. For all elements except the lightest (H, He, C, N, O, Ne) the chemical composition of carbonaceous chondrites is very similar to that of the Sun, suggesting that these meteorites have primitive compositions, were formed from the PSN and have since remained unmodified.

(h)   Compared with solar and chondritic element abundances, the Earth's crust is depleted in core-forming siderophile and chalcophile elements and enriched in crust-and-mantle forming lithophile elements. This leads to the conclusion that the *bulk* composition of the Earth may be chondritic (the chondritic Earth model). The Earth either condensed its core, and then its mantle from chondritic material (heterogeneous accretion); or was originally homogeneous chondritic material that subsequently separated into layers. Like chondrites, the Earth contains low concentrations of all light elements, although oxygen, mainly combined into silicates, is present in almost 'cosmic' proportions.

2.5 Objectives for Section 2

Now that you have completed Section 2, you should be able to:

1 Recognize valid definitions of the terms flagged in the margin of this Section.

2 Outline the main features of the Solar System and recognize some of the physical and chemical factors that are relevant to the composition, origin and evolution of the planets.

3 Describe the basis for determining the composition of the Sun's atmosphere from spectral observations and recognize correct statements about this composition.

4 Recognize examples of nuclear fusion reactions that take place during the lifetimes of stars and account for the main features of the 'cosmic element abundances' curve.

5 Give reasons for the importance of chondritic meteorites in understanding the composition of the Earth and outline the criteria used in selecting the carbonaceous chondrites as the sub-group that most closely represents the composition of non-volatile matter in the PSN.

6 Recognize definitions of the terms lithophile, chalcophile and siderophile and use them in relation to the broad distribution of chemical elements in a layered Earth.

7 Using data concerning the physical and chemical properties of hypothetical and real planets, make broad deductions about their probable compositions and outline how they may have accreted.

Apart from Objective 1, to which they all relate, the six ITQs in this Section test these Objectives as follows: ITQ 1: Objective 2; ITQ 2: Objective 4; ITQ 3: Objectives 3 and 4; ITQ 4: Objective 5; ITQ 5: Objectives 6 and 7; ITQ 6: Objectives 2 and 7.

You should now do the following SAQs, which test other aspects of these Objectives.

SAQ 1 (*Objective 3*)   Look at Figure 12, which gives alkali metal emission spectra for (a) a granite, (b) a basalt and (c) a mineral, mica. What can you say about the *relative* abundances of potassium (K), rubidium (Rb), sodium (Na) and caesium (Cs) in these three samples?

(*SAQ answers begin on p. 96.*)

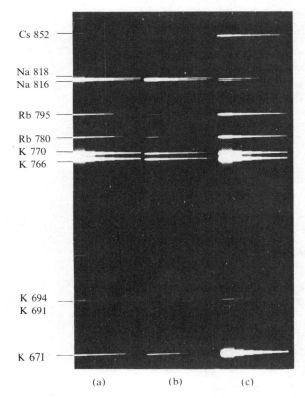

FIGURE 12   Alkali metal emission spectral lines for (a) a granite, (b) a basalt and (c) a mineral, mica. (Line wavelengths are in nm (10^{-9} m).)

SAQ 2 (*Objective 3*) (a) Very strong absorption lines at frequencies of about 4.58×10^{14} and 6.25×10^{14} Hz are found in a stellar spectrum. What elements does this suggest that the star contains, given that these frequencies correspond to 655 nm and 480 nm?

(b) Are the following statements about cosmic element abundances valid or invalid?

(i) Elements of low atomic number are generally more abundant than those of high atomic number.

(ii) Elements of even atomic number are generally less abundant than those of odd atomic number.

SAQ 3 (*Objectives 2 and 5*) Summarize the characteristic features of the principal meteorite types (i–iv) by matching them with the appropriate characteristics (A–G).

(i)	Chondrites	A	Consist essentially of an Fe–Ni alloy
(ii)	Achondrites	B	Dominantly Mg–Fe silicates
(iii)	Stony–irons	C	Contain approximately equal amounts of Fe–Ni alloy and Mg–Fe silicates
(iv)	Irons	D	Contain chondrules
		E	Contain little or no Fe–Ni alloy
		F	Resemble terrestrial igneous rocks
		G	Contain 10–20 per cent Fe–Ni alloy

SAQ 4 (*Objective 6*) Match the possible forms of occurrence of elements (i–vi) with the geochemical classifications (A–C).

(i)	Form silicate minerals	A	Lithophile
(ii)	Occur as sulphides in ore deposits	B	Chalcophile
(iii)	Found as native metals in the Earth's crust	C	Siderophile
(iv)	Occur as oxides in ore deposits		
(v)	May be alloyed with metallic iron in the core		
(vi)	Concentrate in the sulphide phase of meteorites		

SAQ 5 (*Objective 6*) Analyses of gold in chondrites show concentrations of about 1.5 p.p.m. (parts per million) in the metal phase and about 0.005 p.p.m. in the silicate phase. Assume that the relative abundance of gold in the silicate and metal phases of chondrites is the same as in the mantle and core of the Earth. What is

(a) the mass of gold in the Earth's mantle and core?

(b) the proportion of the Earth's gold that is in the core?

(Approximate mass of mantle $= 4 \times 10^{24}$ kg; of core $= 2 \times 10^{24}$ kg.)

SAQ 6 (*Objectives 4, 5 and 7*) Decide, giving reasons, whether each of the following statements is true or false.

(a) Carbonaceous chondrite meteorites have higher concentrations of volatile elements than ordinary chondrites and this is one reason why they are thought to be more primitive.

(b) The most abundant elements with mass numbers between 16 and 56 in solar spectra are the result of processes involving hydrogen burning inside stars.

(c) In order of decreasing abundance, the six most abundant elements in the PSN were hydrogen, helium, carbon, oxygen, nitrogen and silicon.

(d) One of the most common reactions that takes place during a supernova event is that of helium burning to produce carbon.

(e) The detection of short-lived radioactive daughter isotopes in the Solar System indicates that the PSN condensed into the Sun and planets soon after a supernova event.

(f) If the Earth was homogeneous when it accreted, it must have experienced gravitational and/or radioactive heating to cause melting and the development of a layered structure.

(g) The chondritic Earth model implies that the Earth's crust should have a similar composition to the carbonaceous chondrite meteorites.

30

3 Geochemistry, mineralogy and crystal chemistry

Study comment This Section is the core of Block 1; it develops the geochemical classification of the chemical elements, using their electronic properties. This is a classification that describes the occurrence and distribution of elements in nature, particularly their occurrence in the crystalline components of rocks: minerals. Now that you know something about the relative abundances of the chief elements in the Earth (from Section 2), it is natural to enquire into their associated compounds in the Earth. Most of these are silicate minerals, and they are described in Section 3.3.

The major structural subdivisions of the Earth are an outer mantle (plus crust), an outer core and an inner core (Figure 8). As you will find later in the Course, in the crust and mantle, which are dominated by silicates, oxygen is the most abundant element by numbers of atoms.

But, taking the Earth as a whole, which other four elements are also very abundant?

From Figure 8, you might have deduced that iron, magnesium, sulphur and silicon are abundant as well as oxygen. The estimated relative concentrations of these and other elements in a chondritic Earth are shown in Figure 13. The estimated abundances of O and S are more approximate than those for Fe, Si and Mg.

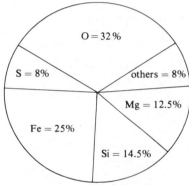

FIGURE 13 Approximate concentrations (weight per cent) of the major chemical constituents in the solid materials of a chondritic Earth. Carbon, hydrogen (in water) and other heavier elements make up the remaining 8 per cent.

From its initial 'nebular' composition, which might have been chondritic as in Figure 13, the layered Earth must have been produced by segregation processes either before or after accretion. We have already hinted that the chemical divisions in the Earth have something to do with the properties of the elements themselves. The same properties are responsible for determining the distribution of all the other elements in the Earth—the 'other' 8 per cent in Figure 13—and so we will start this Section by explaining, in a little more detail, how the elements are *classified geochemically*. As you progress through Section 3 you will find that this classification allows us to understand how and why the elements form the crystalline minerals that we observe in nature.

geochemical classification

3.1 The geochemical classification of the elements

Study comment As you read through Section 3.1 you will find it useful to keep looking at Table 5. If you find chemistry difficult, the next few pages may be tough going; in that case, we suggest that you concentrate on the main points summarized in Section 3.1.3. Table 5 is reprinted on p. 99 and can be cut out for easy reference.

We shall first observe those elements that characteristically enter metal, sulphide and silicate *phases** and then attempt an explanation of the observations. Meteorites provide an excellent source of observations and in Table 4 we record the concentration of 18 elements in these three phases.

phase

*Here the word *phase* is used in its chemical sense to refer to separate parts of a material system (meteorites in Table 4) distinguished by their chemical composition or physical state.

TABLE 4 Distribution of some elements between the metal, sulphide and silicate phases of meteorites. Concentrations are in parts per million by weight.

	metal (nickel–iron)	sulphide (troilite)	silicate
Ca	—	—	16 700
Th	—	—	0.04
Mg	—	—	188 000
Al	—	—	16 000
Ti	—	—	. 800
Si	—(+)	—	225 000
Na	—	—	8 400
Cr	300	1 200	3 900
Fe	907 000	635 000	98 800
Ni	88 000	1 000	100
Pb	0.2	12	1
Cu	200	500	1.5
As	12	10	0.3
Pt	8.4	0.7	1.6
Ru	5.2	0.03	0.3
Os	1.4	—	0.2
Ir	2.8	—	0.5
Au	1.5	0.1	0.005

(—) This signifies that the element is negligible in abundance in that phase as compared with the other phases.

(+) Some Si is found in the metallic phase in meteorites which have been subject to a high degree of reduction (the enstatite chondrites—see Table 2).

ITQ 7   You will see from Table 4 that some elements are (a) strongly lithophile; (b) dominantly chalcophile but with siderophile and/or lithophile tendencies; (c) dominantly siderophile.

To which group does each of the following elements belong?

Cu, Mg, Au, Th, Pb, Os, Ca, Pt.

3.1.1 Periodic correlations

The affinities of any particular element depend very much on the degree of oxidation of the system in question. For example, chromium and manganese are both strongly lithophile in the Earth's crust, but in strongly reduced systems, such as the enstatite-bearing (E-type) chondrites (Table 2 and Figure 9), they form sulphides. These tendencies to cross 'boundaries' between the elements that are lithophile, chalcophile and siderophile are all recorded in Table 5, which is a version of the Periodic Table[A].

Just to remind you about the positions of elements in the Periodic Table, Figure 14 shows the order in which the various *electron sub-shells*[A] increase in energy, and hence in distance from the nucleus. Remember that an electrically neutral atom must have the same number of electrons as protons in the nucleus (the atomic number) and this number is unique for each element. From Figure 14, you can see that the first Period comprises just two elements because there is only one electron sub-shell, and that is of s-type. Periods 2 and 3 have both s and p sub-shells and each comprises eight elements. Subsequent Periods contain more elements because d and f sub-shells are also involved. Elements corresponding to the 3d, 4d and 5d sub-shells (in Periods 4–6) are listed in full in Table 5 and, to make the Table easier to understand, we have reduced to a single square the lanthanide (4f shell, Period 6) and most of the actinide (5f shell, Period 7) groups. (The elements Ac, Th, Pa and U are actinides; the rest are summarized in the square labelled Np–Lr, though Lr is the only known element with a 6d electron.)

You will see in Table 5 that elements of a given affinity (as defined in Section 2.3.2) occupy quite distinct parts of the Periodic Table. For example, the siderophile metals are almost exclusively confined to the transition element groups. Chalcophile tendencies seem also be to related to transition elements and the higher periods of Groups III to VI. Groups I, II, and VII are all lithophile, as are the first few groups of transition elements. Also recorded here is a fourth class of *atmophile*

atmophile elements

TABLE 5 Periodic Table of the elements, showing geochemical tendencies. Electro-negativities appear below element symbols. A heavy line surrounds all elements that show either lithophile or atmophile affinities in the Earth.

GROUP / PERIOD	I	II	TRANSITION ELEMENTS (+ LANTHANIDES + ACTINIDES)										III	IV	V	VI	VII	INERT GASES
1	H 2.1																	He —
2	Li 1.0	Be 1.5											B 2.0	C 2.5	N 3.0	O 3.5	F 4.0	Ne —
3	Na 0.9	Mg 1.2											Al 1.5	Si 1.8	P 2.1	S 2.5	Cl 3.0	Ar —
4	K 0.8	Ca 1.0	Sc 1.3	Ti 1.5	V 1.6	Cr 1.6	Mn 1.5	Fe 1.8	Co 1.8	Ni 1.8	Cu 1.9	Zn 1.6	Ga 1.6	Ge 1.8	As 2.0	Se 2.4	Br 2.8	Kr —
5	Rb 0.8	Sr 1.0	Y 1.2	Zr 1.4	Nb 1.6	Mo 1.8	Tc 1.9	Ru 2.2	Rh 2.2	Pd 2.2	Ag 1.9	Cd 1.7	In 1.7	Sn 1.8	Sb 1.9	Te 2.1	I 2.5	Xe —
6	Cs 0.7	Ba 0.9	La-Lu 1.1–1.2	Hf 1.3	Ta 1.5	W 1.7	Re 1.9	Os 2.2	Ir 2.2	Pt 2.2	Au 2.4	Hg 1.9	Tl 1.8	Pb 1.8	Bi 1.9	Po 2.0	At 2.2	Rn —
7	Fr 0.7	Ra 0.9	Ac 1.1	Th 1.3	Pa 1.5	U 1.7	Np-Lr 1.3											

ATMOPHILE　　LITHOPHILE　　CHALCOPHILE　　SIDEROPHILE

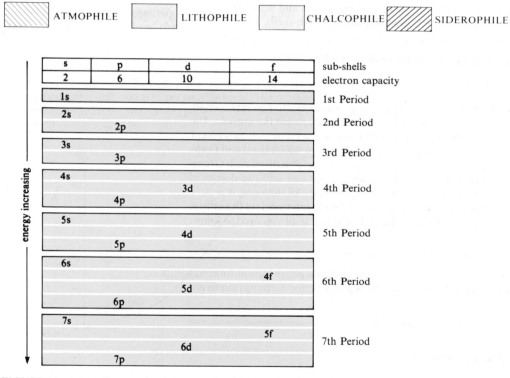

FIGURE 14 Sub-shell energy-level diagram showing the order (increasing downwards) in which sub-shells are filled to complete the electron complement of neutral atoms of different elements. Remember that the number of electrons in neutral atoms must equal the atomic number (the number of protons in the nucleus).

elements, including the chemically inert noble gases, which occur in the Earth's atmosphere. *Clearly, there is a relationship between this geochemical classification and the atomic properties of the elements. The classification must, therefore, also be related to the fundamental structures of the elements.*

Take the example of siderophile elements. These elements have electronic properties that favour the formation of *metallic* bonds (see Section 3.1.2 for further details); they do not readily form bonds with oxygen or sulphur. A basis for describing the ability of atoms to form different kinds of bonds can be derived from their

From the reasoning in ITQ 11(b), it follows that the higher the positive charge, the smaller will be the cation. For example, manganese can have the following ionic radii:

$$Mn^{2+} \qquad 80 \, pm$$
$$Mn^{3+} \qquad 66 \, pm$$
$$Mn^{4+} \qquad 60 \, pm$$

So the charge of a multivalent atom may determine the co-ordination number for that atom through its effect on ionic size.

We have demonstrated already (for example, in Figure 13) that oxygen and silicon are likely to be the two most abundant elements in the Earth's crust and mantle. If you refer to Figure 21, you can see that oxygen forms the *largest* common ions that occur in silicate minerals and so the structures of these minerals might well be regarded as closely packed clusters of oxygen ions, with cations filling in the interstices, according to the rules of co-ordination (Table 7).

> **ITQ 12** On the basis of radius ratios, what type of co-ordination would you expect to find between oxygen and (a) Si^{4+}, (b) Rb^+, (c) Fe^{2+}, (d) Sr^{2+}, (e) B^{3+}?

If we compare predictions about co-ordination, such as those in ITQ 12, with observed co-ordination in minerals, we find a good agreement except for some overlap at the boundaries defined in Table 7. These boundaries move a little as pressure and temperature vary: in particular, high pressures favour higher co-ordination numbers for any particular cation. For example, the most important basic structural unit in silicate minerals is the anion complex $(SiO_4)^{4-}$ which has tetrahedral (4-fold) co-ordination of silicon by oxygen (Figure 22). But at very high pressures, a variety of SiO_2 (normally tetrahedral) is produced with octahedral (6-fold) co-ordination. This is the mineral *stishovite*, which has a much greater density than normal SiO_2 (quartz) because more oxygen ions are squashed around each silicon ion. Stishovite is found in meteorite impact craters and may also occur at the high pressures which exist deep in the mantle. Further details of increases in co-ordination numbers at high pressures are discussed in Block 2 in relation to the mantle transition zone. Note that ionic radii vary slightly depending on the co-ordination number and, strictly, the values quoted in Figure 21 are for octahedral co-ordination only. However, deductions of the sort made in ITQ 12 are generally valid for studying silicate minerals of the Earth's crust.

stishovite

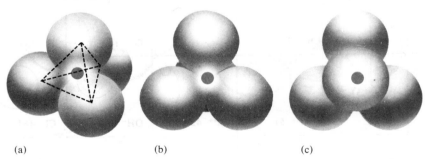

(a) (b) (c)

FIGURE 22 Three views of an SiO_4 tetrahedron. The large spheres are oxygen and the small red sphere in the centre is silicon.

As you will find in Section 3.3, most silicate minerals contain several different elements which, according to the rules used in ITQ 12, have different co-ordination numbers. As we have said, tetrahedra by themselves cannot be stacked together to fill space and build structures. What happens is that silicate tetrahedra are linked together by other cations that have larger co-ordination numbers with oxygen (as, for example, Fe^{2+} or Mg^{2+}, which occur in octahedral co-ordination in olivines). The other complication is that most silicate minerals do not comprise the simple $(SiO_4)^{4-}$ tetrahedra of Figure 22, but are made from linked tetrahedra in which *oxygen sharing*[A] occurs. This idea was illustrated in the Science Foundation Course and will be developed fully in AV 01 and Section 3.3. Basically, what happens is that as progressively more oxygens are shared between adjacent tetrahedra, the net negative charge on the anion complex is reduced. The different ways in which tetrahedra are linked through their oxygens to give giant anions, or *polytetrahedra*, explains the natural diversity of silicate minerals. But before we look at *silicate polytetrahedra* in more detail there are two other aspects of crystalline behaviour about which you should know: isomorphism and polymorphism.

silicate polytetrahedra

3.2.2 Isomorphism and solid solution

Two minerals with different chemical compositions but with closely related chemical formulae and *identical atomic structures* are said to exhibit *isomorphism*. This term, meaning literally 'equal form', was useful to the early chemists in their attempts to determine valencies and atomic weights from chemical reactions. Many examples occur in silicate systems; for example, in the olivine, pyroxene and feldspar groups (which will be discussed in Section 3.3). For isomorphism to occur, anions and cations of the same *relative* size and in the same number must crystallize in the same structure.

isomorphism

Although some substances that are isomorphous have structures that are very different in size, there are others where the structures are of similar size, as well as shape. In the latter case, there may be *partial or complete ionic substitution* of one element for another where the atoms of the two elements form ions of similar radius. One of the best-known examples of complete ionic substitution occurs in the olivine mineral group, where all compositional variants occur between the two isomorphous *end-members* that you have already met in ITQ 10 and Figure 18: Fe_2SiO_4 (fayalite) and Mg_2SiO_4 (forsterite). In this case, Fe^{2+} and Mg^{2+} substitute for each other in the mineral structure because they are so nearly equal in ionic size that either may be accommodated in octahedral *co-ordination sites* (the available holes in the centre of octahedral co-ordination polyhedra) without the structure of the mineral being significantly disrupted. In the olivine minerals, where complete ionic substitution is possible between two end-members, a complete range of 'mixed crystals' may form. We refer to these mixed crystals resulting from complete ionic substitution as members of a *solid solution series*. To indicate this property, the chemical formula of olivines is often expressed as $(Mg,Fe)_2SiO_4$ to indicate the complete substitution between Fe^{2+} and Mg^{2+}.

ionic substitution

end-members

co-ordination sites

solid solution series

Now let us consider what happens when Ca^{2+} tries to enter the olivine structure. Calcium forms a much larger ion (radius 99 pm) and cannot be accommodated so easily in the structure of olivine; consequently, only very *limited ionic substitution* of Ca^{2+} for Fe^{2+} or Mg^{2+} occurs. Thus there is a very useful general rule:

One ion may replace another completely, to form a solid solution series, if the difference in their ionic radii does not exceed 15 per cent of the radius of the smaller ion. If their size difference is greater than this but both ions are still capable of occupying the same site (as defined by Table 7), then *partial ionic substitution* will occur.

ITQ 13 Aluminium is the third most abundant element in the Earth's crust, after silicon and oxygen. On the one hand, it forms the *complex aluminate anion* $(AlO_4)^{5-}$ which is analogous to the silicate anion $(SiO_4)^{4-}$ and, on the other hand, it can also adopt octahedral co-ordination in silicates in the same way as Fe^{2+} and Mg^{2+} do.

(a)   Using Table 7 and the ionic radii in Figure 21, can you explain this behaviour in terms of co-ordination number?

(b)   Can Al^{3+} substitute for either Mg^{2+} or Fe^{2+} in the same structure?

(c)   There are silicates in which Al^{3+} substitutes for Si^{4+}. Do you think this will be *partial* or *complete* ionic substitution?

The answer to ITQ 13(c) gives an example of the fact that ions differing in size by more than 15 per cent may be involved in partial ionic substitution. One of the most important examples of partial substitution of Al^{3+} for Si^{4+} occurs in the plagioclase feldspar mineral group (which will be discussed in Section 3.3) where the end-members, $NaAlSi_3O_8$ and $CaAl_2Si_2O_8$, contain 25 and 50 per cent respectively of Al^{3+} in tetrahedral co-ordination sites.

Do you think there will be a solid solution series between these two end-members?

Given that we can make the necessary Al^{3+}–Si^{4+} substitutions, we need only to consider the size difference between Na^+ and Ca^{2+} ions. From Figure 21 this is 2/97, or just over 2 per cent, so there is complete ionic substitution between Na^+ and Ca^{2+}: *the plagioclase feldspars are another example of a solid solution series.*

There is another interesting point about these feldspars. They demonstrate that ionic size is more important than ionic charge in forming solid solution series because some charge inbalance can always be compensated for by substitutions elsewhere in the structure. The formulae of the feldspars (given above) show that between 1 in 4 and 2 in 4 of the silicons in the tetrahedral sites can be replaced in going from $NaAlSi_3O_8$ to $CaAl_2Si_2O_8$. But the substitution of Al^{3+} for Si^{4+} means that there is an extra negative charge in the structure that has to be balanced. That is done by the substitution of Ca^{2+} for Na^+. In other words, the substitution of CaAl for NaSi keeps the structure electrically balanced. This is known as *coupled substitution*. With partial and complete ionic substitution possible on this scale in silicates, it is not surprising that the early geochemists could not classify these minerals by their compositions alone.

coupled substitution

As we implied earlier, you should not equate isomorphism and solid solution behaviour because many isomorphous substances have little or no ionic substitution: $CaCO_3$ (calcite) and $ZnCO_3$ (smithsonite), for example. As you can see from Figure 21, the ionic radii of zinc and calcium are sufficiently similar for them to occupy similar co-ordination sites but they differ by much more than the permitted amount (in fact, by 25/74 or 34 per cent) for complete substitution to occur. So the distances between the Zn^{2+} cations and CO_3^{2-} anion complexes in smithsonite are smaller than those in calcite. Although the two structures are identical in shape, a molecule of $CaCO_3$ is bigger than a molecule of $ZnCO_3$.

To check your understanding of the main points in this important Section, try the following ITQ.

> **ITQ 14** (a) The salts KCl and KBr form a complete range of solid solutions but the salts NaCl and KCl have only limited ionic substitution between them. Can you explain the difference?
>
> (b) A second group of feldspars, known as the alkali feldspars, has the end-members $NaAlSi_3O_8$ and $KAlSi_3O_8$. Bearing in mind your answer to (a), would you expect complete or partial ionic substitution in the alkali feldspars?

3.2.3 Polymorphism

In *polymorphism*, a single element or compound may occur in more than one crystal form with different packing arrangements of its constituent atoms, ions or molecules. Therefore, polymorphism, which literally means 'in many forms', is a different condition from isomorphism. Two or more minerals with the same chemical composition but contrasting structures are *polymorphs*.

polymorphism

> Earlier we mentioned two polymorphs of silica, SiO_2. Can you remember their relationship?

Here we are referring to the change in Si—O co-ordination from tetrahedral to octahedral with increasing pressure. The low-pressure form, *quartz*, has a density of $2\,650\,kg\,m^{-3}$, whereas that of the high-pressure polymorph, *stishovite*, is $4\,300\,kg\,m^{-3}$.

There are some other polymorphs of SiO_2 that are stable at different conditions of pressure and temperature. One that forms at high temperatures is *tridymite*. It has a less closely packed structure than quartz, its density is $2\,260\,kg\,m^{-3}$, and it is found in volcanic rocks that have cooled rapidly from temperatures above 850 °C. The precise details of these polymorphic structures do not need to concern us, but you should realize that it is possible to determine conditions of polymorphic transitions by experiments and to use the results to study the pressure and temperature under which a rock has formed.

> Stishovite forms in high-pressure meteorite impact events. How is it then that we can find stishovite at the Earth's surface under ordinary atmospheric pressure long after such events? In other words, why has the stishovite not reverted to the low-pressure form, quartz?

The answer is that some polymorphic transitions are more readily reversible than others. The quartz ⟶ tridymite transition is reversed with *slow* cooling at

867 °C and 1 atmosphere pressure* but other transitions, such as that producing stishovite, are less easily reversed. When studied in detail, over a wide range of pressures and temperatures, most polymorphic transitions can be reversed. A simple analogy might help you to understand this. In Figure 23, a ball is placed on a wooden block with a ledge and a lower step, which represents a lower energy level. On the higher level, the ball is less stable than on the lower level but it will not leave the higher level unless an external influence, such as someone bumping into the block, provides the tiny triggering quantity of energy needed for the ball to roll over the ledge and to fall to the step. So you can imagine that the right-hand side in Figure 23 represents stishovite at low pressure and the left-hand side quartz. Stishovite will persist, even though it is the less stable (or *metastable*) form of silica, unless we find some way of providing the triggering energy, such as heating at low pressure, that will make it revert to the more stable form. The height of the ledge in Figure 23, or energy barrier, that must be overcome varies depending on the polymorphs involved, and you can think of rapidly reversing transitions as having small energy barriers. Note that the uphill process in Figure 23 will never occur unless the triggering energy exceeds the energy needed for the uphill step. So the size of the step is important, and we will come back to this analogy shortly.

Another example of a metastable polymorph in everyday experience is the mineral *diamond*, a polymorph of carbon. It has long been known that natural diamonds can be reduced to a valueless deposit of black graphite dust by strong physical shock but, until the 1950s, the opposite reaction (graphite ⟶ diamond) had not been observed. Equipment capable of generating pressures up to 200 kilobars has now been developed and this has allowed many chemical reactions that take place in the outer 600 km of the Earth to be studied directly. On the commercial front, it has allowed large quantities of industrial diamonds to be manufactured. The polymorphic transition boundary for graphite to diamond is shown in Figure 24. The *stability fields* labelled diamond and graphite represent 'pressure–temperature space' (that is, areas in the graph) where the polymorph indicated is the more stable form of carbon.

Now go back to our analogue model in Figure 23 and, this time, call the left-hand side graphite and the right-hand side diamond.

> Does this represent the situation above or below the transition boundary in Figure 24?

You should have argued that, because graphite has the lower energy level in Figure 23, it is the more stable polymorph, and thus this situation occurs at low pressures, *below* the line in Figure 24. But at high pressures, diamond must be the more stable (lower energy level) polymorph and the situation is reversed.

> Given that the temperature in the Earth's upper mantle, below 100 km depth, is about 1 500 °C, can you estimate from Figure 24 the minimum pressure and depth needed for diamonds to be produced?

The minimum pressure is just over 40 kbar and this corresponds to a depth of about 150 km. This raises a very interesting point that we shall take up again in Block 2: all naturally occurring diamonds (non-industrial) must have come from at least 150 km depth in the Earth. They are thought to have risen from this depth, together with their host rocks, through volcanic 'pipes'.

The fact that diamond is the high-pressure polymorph of carbon can also be predicted from a knowledge of the internal structures of graphite and diamond (Figure 25). These polymorphs provide an excellent example of the way in which the physical properties of minerals are a function of their internal structures. In graphite, the carbon atoms are arranged in sheets: the bond lengths between adjacent atoms within sheets are 142 pm, whereas those between the sheets are 340 pm. Compare this with diamond, where the carbon–carbon distances are all 154 pm, resulting in extremely strong bonding. Perhaps you can see why diamond is very much harder than graphite, which breaks, or *cleaves*, between the sheets.

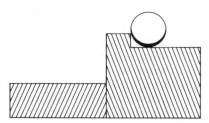

FIGURE 23   Wooden block model of energy levels (see text for details).

graphite–diamond polymorphism

stability fields

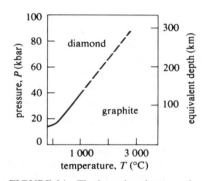

FIGURE 24   The boundary between the stability fields of graphite and diamond in pressure–temperature space.

kilobars

* A word about pressure units: *kilobars* (kbar) are used frequently by geologists. 200 kilobars is 200×10^3 bars, where 1 bar is about equal to pressure of 1 atmosphere. In SI units, 1 bar $\equiv 10^5 \, N \, m^{-2}$, or 10^5 pascals, so 200 kilobars is $2 \times 10^{10} \, N \, m^{-2}$ or $20 \, GN \, m^{-2}$. (This is the pressure that would result if about 1 million elephants, of average mass, were standing on 1 square metre—and that is the pressure found at 600 km depth in the Earth!)

This is the ideal point for you to complete the Audio-vision/Home Experiment sequence AV 01: 'Mineral structures', in which we ask you to make some models of mineral structures using your Home Experiment Kit. During this sequence you will experiment with the different ways in which silicate tetrahedra are joined and we shall also tell you about the way in which ions and ionic complexes are packed together to form mineral structures. Try to keep the distinction in your mind between *co-ordination polyhedra* (Figures 19 and 20) and *silicate polytetrahedra* (based on multiples of the structure shown in Figure 22). The former describe the way individual atoms are co-ordinated in atomic structures, whereas the latter describe the linking of silicate tetrahedra, through their oxygens, to form massive anion complexes. The AV sequence will take you about an hour to complete and it starts in the AV Notes, to which you should now refer.

When you have completed the AV sequence, try the following ITQ:

> **ITQ 16**   In terms of their silicon and oxygen contents, what is the simplest unit formula and net charge of each of the following silicate anion complexes:
>
> (a)   a single silicate tetrahedron (Plate 7)?
>
> (b)   a one-dimensional row of linked tetrahedra (Plate 9)?
>
> (c)   a double row of linked tetrahedra (Plate 10)?
>
> (d)   a two-dimensional array, or sheet of tetrahedra (Plate 11)?
>
> (e)   a complete three-dimensional framework with silicon and oxygen only (Plate 18)?

You should understand that these negatively charged tetrahedra and polytetrahedra, which form the skeletons of all silicate mineral structures, are held together and are electrically neutralized by positively charged metal cations, to make very stable configurations. *The arrangement of these cations in crystalline structures is determined both by their ionic radii and by their valency state which, in turn, depends on their electronic configurations.*

In the following Sections, we examine the ways in which the silicate complexes and available cations are bonded together to form different structures. We shall confine our attention to the most important examples of silicate minerals, starting with the simplest anion complexes, and build up to the largest 'framework' structures at the end.

Most of the information that we should like you to glean from Section 3.3 is summarized in Table 8. In this case, we are breaking the tradition of earlier Sections and giving you this summary *before* the main evidence because it will help to focus your ideas as you read the following text. Bear in mind two things as you proceed:

(i)   the structural and chemical complexity of the silicate minerals increases *down* Table 8; in a simple way, the sequence is towards an increasing ratio of silica to cations;

(ii)   silicate minerals make up more than 99 per cent of the Earth's crust and mantle and so it is important that you understand something about their structure, composition and properties. What is more, they are 'assumed knowledge' in the later parts of this Course. Spare a few minutes to take a closer look at Table 8, therefore, before you read any further. There is no need to try to *remember* all the mineral formulae given in Section 3.3; examples of each group, which you should be able to recall, are given in Table 8.

3.3.1   Separate tetrahedra—olivines and garnets

There are two important sub-groups of silicate minerals whose structures are based on the simplest anion group, $(SiO_4)^{4-}$: the olivines and the garnets (*CB*, Plates 12 and 13). Both are semi-precious gem stones and may be familiar to you (gem olivine is called peridot). *Olivines* are named after their colour, olive green, which is due to the presence of reduced iron (Fe^{2+}) in the structure of the mineral. They form one of

olivines

TABLE 8 Summary of the atomic structure and other properties of the main groups of silicate minerals

Structural type and silicate unit	Mineral types with examples of formulae	Additional comments	Density $(kg\,m^{-3})$
Single tetrahedra $(SiO_4)^{4-}$	Olivines: Mg_2SiO_4 Garnets: $Mg_3Al_2(SiO_4)_3$	Dense compact structures. Crystals have equidimensional shapes but no cleavage. (Olivine structure is shown in Figure 28.)	3 200–3 600
Single chains $(Si_2O_6)^{4-}$	Pyroxenes: $Mg_2Si_2O_6$	Pyroxenes are generally equidimensional and dense, but they have two well-developed cleavages (nearly at 90°) because cation bonding is weaker between chain bases than vertices (Figure 30).	3 200–3 600
Double chains $(Si_4O_{11}(OH))^{7-}$	Amphiboles: $Ca_2Mg_5(Si_8O_{22})(OH)_2$	Similarly shaped, but less compact and less dense than pyroxenes because there are holes in the double chains; cation bonding is generally weaker than in pyroxenes, though there is a similar internal arrangement (Figures 31 and 32): two cleavages at 56°. The holes in the double chains accommodate $(OH)^-$ groups.	2 800–3 200
Sheet silicates $(Si_4O_{10})^{4-}$	Talc: $Mg_3Si_4O_{10}(OH)_2$ Micas: $KMg_3AlSi_3O_{10}(OH)_2$ (also clay minerals)	Plate-like minerals with open structures and low densities. The sheets are not bonded to each other electrically in talc and are only bonded by weak K—O (12-fold co-ordination) bonds in mica (Figure 35). Although the sheets themselves are strong, there is one very well-defined cleavage between them. The open spaces in the sheets accommodate $(OH)^-$ groups.	2 500–3 300
Framework silicates (SiO_2)	Quartz: SiO_2 Feldspar: $NaAlSi_3O_8$	Quartz has a rigid framework with strong Si—O bonds only. Feldspars have some weaker bonds (between alkali metal atoms and oxygen) and show two poorly-developed cleavages, nearly at 90°. Although their structures are more compact than micas, for example, neither mineral group contains heavy elements and so densities are low.	2 600–2 800

the best known examples of a solid solution series in the silicates, between Fe_2SiO_4 and Mg_2SiO_4 (Figure 18), in which Fe^{2+} and Mg^{2+} ions are completely interchangeable because of their closely similar radii (Figure 21).

Although the most common olivines are iron- and magnesium-rich, there are some less common varieties that contain Ca^{2+} or Mn^{2+} as their important cation. These olivines have similar crystal structures to those of common olivines but will they enter into solid solutions with common olivine? (*Hint* You will get different answers for Ca^{2+} and Mn^{2+}.)

The relevant ionic radii are Ca^{2+}, 99 pm; Mn^{2+}, 80 pm; Fe^{2+}, 74 pm; Mg^{2+}, 66 pm. The *calcium* ion is so much larger than either iron or magnesium ions (34 and 50 per cent larger respectively) that there is not even likely to be much partial ionic substitution of Ca^{2+} for Fe^{2+} or Mg^{2+}. There is no question of a solid solution series (requiring complete ionic substitution) involving calcium in olivines. But the occurrence of calcium-rich olivines with the same crystal structure as common olivines provides another example of *isomorphism* without solid solution. *Manganese* ions are more closely similar in size to iron and magnesium ions (the difference in size is 8 and 21 per cent respectively). We should expect, therefore, complete ionic substitution of Mn^{2+} in Fe_2SiO_4, but only partial ionic substitution in Mg_2SiO_4. So manganese enters into solid solutions with iron-rich olivines; generally, however, Mn^{2+} is less abundant than Fe^{2+} in rocks and the proportion of Mn_2SiO_4 in olivine is usually very small.

The atomic structure of magnesium olivine, *forsterite*, is shown in Figure 28. Please read the caption carefully as this explains how to interpret the picture.

forsterite

How many complete $(SiO_4)^{4-}$ groups (i.e. whole or part $(SiO_4)^{4-}$ groups that together constitute whole numbers of groups) and how many Mg^{2+} cations are shown in Figure 28? Does this suggest that there is a particular ratio of cations to silicate units in olivines?

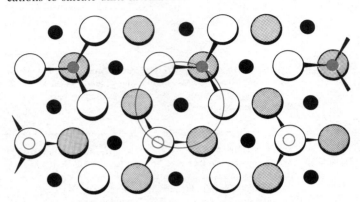

FIGURE 28   Two-dimensional projection of the atomic structure of magnesium olivine, forsterite. The large circles represent oxygen atoms in tetrahedral arrangement around silicon atoms (small red circles), and the black circles are magnesium cations. Notice that alternate rows of separate tetrahedra point up and down, giving an upper plane (shown unshaded) and a lower plane (shown with grey shading) of oxygen atoms; between the two planes of oxygen atoms lie most of the magnesium cations (black). The red ring identifies the oxygen atoms that are associated with one central magnesium ion and shows that the magnesium ions are in octahedral co-ordination with oxygen (see Figure 20c).

You should find *six* almost complete $(SiO_4)^{4-}$ groups, with some oxygens from another four, and *thirteen* magnesium cations. This is approximately a 2:1 ratio of Mg^{2+} to $(SiO_4)^{4-}$ and we hope you are satisfied that, if we continued the structure in three dimensions to infinity, we should end up with a 2:1 ratio: hence the formula Mg_2SiO_4. Later on, in Section 3.3.5, we shall consider the chemical composition of olivines and other minerals and shall need to use this ratio of cations to silicate units. Of course, all the magnesium atoms in Figure 28 could be substituted by iron atoms to give Fe_2SiO_4, or *fayalite*.

fayalite

Garnets (*CB*, Plate 13) are often red due to the presence of some Fe^{3+}. They have more complicated structures than olivines, as you may appreciate from their general chemical formula, $M_3^{2+} M_2^{3+} (SiO_4)_3$, where M^{2+} are *divalent* cations in cubic co-ordination sites (Figure 20d) and M^{3+} are *trivalent* cations in octahedral sites (Figure 20c). The divalent cations are usually a combination of Fe^{2+}, Mg^{2+}, Ca^{2+} and Mn^{2+}, whereas the trivalent cations are Al^{3+} or Fe^{3+}. Let us see if that makes sense in terms of ionic radii and co-ordination number, especially as in olivines (Figure 28) the divalent ions are in octahedral sites.

garnets

Compare the sizes of the divalent and trivalent cations in garnets, using data from Figure 21.

The divalent cations range in radius from 66 to 99 pm (Mg^{2+} and Ca^{2+}), whereas the trivalent cations are *smaller*, 51 pm (Al^{3+}) and 64 pm (Fe^{3+}). So we should expect the trivalent cations to occupy the smaller spaces available in octahedral co-ordination sites (Figure 20) and that is just what happens. But when you work out the radius ratios for these six cations relative to oxygen, you find they are all less than 0.732. That corresponds to 6-fold co-ordination (Table 7), so all these cations *could* fit into octahedral sites. The point we would like to emphasize is that *cations may sometimes fit into larger sites than you would predict from Table 7* and this is just what happens to divalent cations in garnets. This also helps to explain the more varied chemical compositions of garnets, compared with olivines: larger cations (e.g. Ca^{2+}) can be accommodated easily in cubic co-ordination sites.

Even so, garnets have compact structures and, like olivines, usually contain a significant number of heavy iron atoms. Because of these features, both garnets and olivines have *higher densities* than most other silicates, usually in excess of $3\,200\,kg\,m^{-3}$, compared with c. $2\,650\,kg\,m^{-3}$ for quartz (SiO_2)—see Table 8. You may recall from Le Chatelier's principle that high-density minerals are favoured by high pressures and it is thought that both olivines and garnets form important components of deep crustal and mantle rocks.

That completes our brief look at minerals based on separate silicate tetrahedral units. All the remaining silicate structures we shall consider (Sections 3.3.2–3.3.4) contain giant anion complexes (chains, sheets and frameworks) that can extend indefinitely in one, two and three dimensions.

3.3.2 Chain silicates—pyroxenes and amphiboles

Our starting point for introducing the *chain silicates* is the structural model you made in AV 01, which is shown in *CB*, Plate 9. Each silicate tetrahedron is joined to two others and alternate tetrahedra point outwards from the chain in opposite directions. This is illustrated more simply by approximating each tetrahedron to a triangle, as in Figure 29. (If you are not convinced, have another go at making a single chain from balls and spokes.) Along the chain, the configuration of silicate tetrahedra repeats itself after each pair. An example of such a *repeat unit* is delimited by an arrow in Figure 29. To obtain a manageable chemical formula for chains of indefinite extent, we take the formula of the *smallest* possible representative unit—the repeat unit.

chain silicates

repeat unit

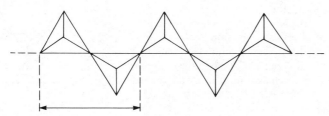

How many silicon and oxygen atoms occur in a repeat unit?

FIGURE 29 Arrangement of the SiO_4 tetrahedra in single-chain silicates (arrow denotes the repeat unit).

Each tetrahedron shares two oxygens with its neighbours, so there are two silicons and six oxygen atoms. *The chemical formula assigned to the silicate skeleton of single chain silicates* is, therefore, $(Si_2O_6)^{4-}$, and this unit is the basis of the *pyroxene* group of minerals. Most pyroxenes are dark green or dark brown like the augite samples shown in *CB*, Plate 14. The chains are held together by Mg^{2+} and Fe^{2+} ions, mainly in octahedral co-ordination sites. As in olivines, there is a complete solid solution series between $Mg_2Si_2O_6$ and $Fe_2Si_2O_6$. There is also some Ca^{2+} in most pyroxene structures which become a little distorted by the large calcium ion.*

pyroxenes

To appreciate how the silicate chains are joined together by these cations take a look at Figure 30. In Figure 30a, at the top, the chains are viewed from above (as in Figure 29) and the projection (below) shows the end-on view of a chain. Successive pairs of

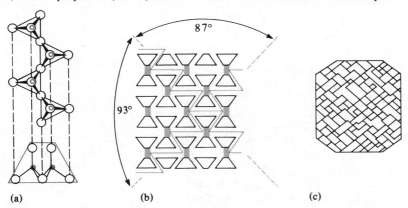

(a)   (b)   (c)

FIGURE 30 The internal atomic structures of single-chain silicates, pyroxenes. (a) shows a chain of linked silicate tetrahedra viewed from above (top) and from the side (bottom). The red outline shows the shorthand symbol used in (b) (in black) to indicate cross-sections of the chains of tetrahedra. (b) shows the stacking of the chains in the crystal structure and the most important sites (red rectangles) for chain-linking cations. The cleavages at 87° and 93° seen in crystals (c) cut at right angles to the chain length are controlled by the way the chains break apart (along the grey lines in (b)).

tetrahedra along the chain are now immediately behind each other and the red quadrilateral drawn around the front pair provides us with a short-hand way of denoting each chain in the structural model (Figure 30b). Here we are looking at many more chains end-on and you should notice two important points:

(i)   the chains are arranged in parallel rows along which alternate chains 'point' in opposite directions;

(ii)   the cations (mainly in octahedral sites) are concentrated in the red rectangles where adjacent chains point towards each other.

* Pyroxenes with near to the natural maximum limit of 50 per cent replacement of Fe^{2+} and Mg^{2+} by Ca^{2+} are called *clinopyroxenes*, whereas those with less Ca^{2+} are called *orthopyroxenes*. You will find further details of these in Block 3.

Because the cations are concentrated between chain apices, the pyroxene structure is only weakly bonded between the chain bases, where it tends to break apart easily, denoted by the grey lines in Figure 30b. These lines join together by running diagonally between the chains. They define the *cleavage* in pyroxenes, planes of weakness along which the crystals break. In crystals of pyroxene, seen end-on, very fine diagonal lines can be observed, intersecting almost at right angles. These are the traces of the cleavage planes (Figure 30c).

cleavage

CB, Plate 15 is a photomicrograph (a magnified thin section photograph) of a pyroxene crystal cut at right angles to the length of the chains. The fine cracks are cleavage traces and it is easy to see how the internal structure of pyroxene crystals controls the *shape* of the crystals (Figure 30c).

In the same way that single chains are linked together to form pyroxenes, so *double chains* (*CB*, Plate 10) may be linked to form another mineral group similar in colour, appearance and structure: the *amphiboles*. The double chains are illustrated alone in Figure 31 and again the repeat unit is indicated with an arrow. This time you can see four tetrahedra in the repeat unit, two pointing outwards from the double chain and two pointing inwards. The latter pair share the oxygen atom that links the two single chains together and so there are eleven oxygen atoms for every four silicon atoms, giving *a chemical formula for the double chains* of $(Si_4O_{11})^{6-}$: all amphiboles have this anion formula. The double chains are stacked into the mineral structure in exactly the same way as pyroxene chains. The only difference is in the shape of the chains in cross-section (Figure 32a); the amphibole chains are wider and this influences the cleavage directions in the mineral (Figures 32b and 32c) which are at angles of 56° and 124°. A photomicrograph of an amphibole cut at right angles to the chain length (*CB*, Plate 16) shows these cleavage directions to good effect.

amphiboles

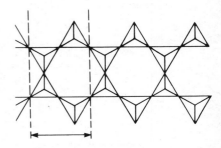

FIGURE 31 Arrangement of the SiO_4 tetrahedra in double-chain silicates (arrow denotes the repeat unit).

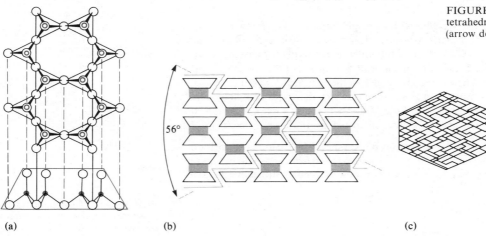

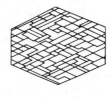

(a)　　　　　　　　(b)　　　　　　　　(c)

FIGURE 32 The internal atomic structures of double-chain silicates, amphiboles. (a) shows a double chain of linked silicate tetrahedra viewed from above (top) and from the side (bottom). The red outline shows the shorthand symbol used in (b) (in black) to indicate cross-sections of the chains of tetrahedra. Compare the shape of the pyroxene chains in Figure 30a. (b) shows the stacking of the chains in the crystal structure and the most important sites (red rectangles) for chain-linking cations. The cleavages at 56° seen in crystals (c) cut at right angles to the chain length are controlled by the way the chains break apart (along the grey lines in (b)).

As in pyroxenes, most of the cations sit between the apices of adjacent chains (the red rectangles in Figure 32b) and again these are mainly Fe^{2+} and Mg^{2+}. But the sizes of the spaces available for cations are larger in amphiboles and so these minerals contain a much greater diversity of larger cations, such as Ca^{2+}, Na^+ in cubic sites, Fe^{3+} and Al^{3+} in octahedral sites, and Al^{3+} in tetrahedral sites. A typical amphibole, known as *tremolite*, has the simplified (!) chemical formula, $Ca_2Mg_5(Si_8O_{22})(OH)_2$.

You may wonder where the $(OH)^-$ groups have come from. There is one hydroxyl group per repeat unit: $[Si_4O_{11}(OH)]^{7-}$ and this is common to all amphiboles.

> With reference to Figure 31, does this suggest a possible *site* for $(OH)^-$ groups in the amphibole structure?

The $(OH)^-$ groups sit right inside the double chains, occupying the hexagonal-shaped spaces that you can see in Figure 31. Now, if you look again at Figure 32b, you should appreciate that *each* chain cross-section contains an $(OH)^-$ site (see also Figure 34). The sites in adjacent chains line up above each other in the structure to form 'holes' that can be thought of as 'drainpipes'. The holes are running from top to bottom in Figure 32b.

> What implications will the presence of $(OH)^-$ groups in amphiboles have for the relative densities of pyroxenes and amphiboles?

$(OH)^-$ groups are of low mass and so *amphiboles will be less dense than pyroxenes* (see Table 8 for a range of values). It is also true that pyroxenes, in general, have more tightly packed and chemically simple structures than amphiboles and this reinforces our conclusion about density.

> **ITQ 17** How does the ratio of the number of silicon atoms to the number of cations in single and double chain silicates compare with the ratio you determined earlier for olivines? (You can use the chemical formulae in Table 8.) If we had chemical analyses of each mineral, which would contain the highest and which the lowest overall proportion of silicon?

Before continuing there are three important points we should like to emphasize:

1 We have used ball-and-spoke models in our diagrams to show relationships between atoms; a more accurate representation would show all the atoms as large balls in contact with each other.

2 In Section 3.2 we were considering the mineral inter-layer spacings that are determined by X-ray diffraction and then used to model atomic structures. To give you some feeling for the meaning of these spacings, the layers of chains and the prominent cleavage directions, shown in Figures 30b and 32b, would give strong reflections: their spacings, roughly 10^3 pm, have all been determined by X-ray diffraction methods.

3 *The three main groups of dense minerals, olivines, pyroxenes and amphiboles, are the main rock-forming minerals that contain Fe^{2+} and Mg^{2+} as important cations. Together with biotite mica (see below), they are known as the* ferromagnesian minerals.

ferromagnesian minerals

3.3.3 Sheet silicates—micas and clay minerals

The next step is for silicate polytetrahedra to form *sheets* that are joined up in two dimensions (Figure 33 and *CB*, Plate 11).

sheet silicates

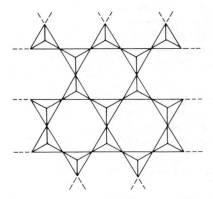

FIGURE 33   Arrangement of the SiO_4 tetrahedra in sheet silicates.

> The repeat unit is usually taken to include four tetrahedra (as in amphiboles). How many oxygen atoms will be associated with each repeat unit?

Three out of every four oxygens around each silicon atom are shared, so there are 2.5 oxygens for each silicon, giving $(Si_4O_{10})^{4-}$ as the formula for polytetrahedra of sheet silicates.

As with amphiboles, you can see that there are rings in the sheet silicate structure which, again, are occupied by hydroxyl groups. To make this clear, look at Figure 34, which gives a view down on top of the sheet with the oxygen anions and hydroxyl anion complexes drawn to their correct scale. We have omitted silicon atoms and the cations for the moment so that you can identify the $(OH)^-$ sites in each ring of the sheet. Because the sheets extend indefinitely in two dimensions, there are twice the number of $(OH)^-$ sites available that there are in amphiboles. So the formula of the complete silicon–oxygen–hydroxyl anion complex shown in Figure 34 is $[(Si_4O_{10})(OH)_2]^{6-}$.

Bonding in sheet silicates is really very simple: if we take a cross-section through the sheets, they are stacked as shown in Figure 35. The sheets are stacked with their vertices pointing towards and facing away from each other alternately. Between the two sets of vertices there are octahedral cation sites where octahedrally co-ordinated cations, usually Mg^{2+}, Fe^{2+}, or Al^{3+}, bind the sheets together.

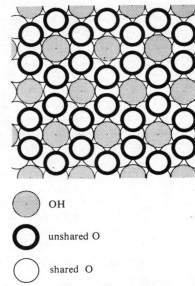

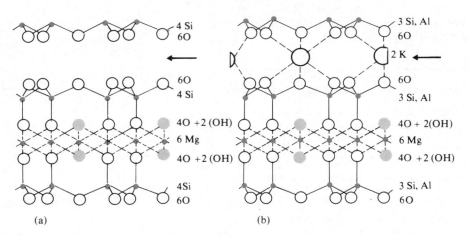

(a)   (b)

FIGURE 35   Two-dimensional projections of cross-sections through the structures of two sheet silicates cut across the sheets: (a) talc, (b) biotite mica. The atoms along each row are labelled; dashed black lines indicate the co-ordination 'bonds' around each cation. (*Note* that in Figure 35b, only the forward-pointing co-ordination bonds around potassium atoms are shown, the other half point backwards into the paper. Potassium is in 12-fold co-ordination.) Arrows refer to planes of weakness in the structure.

FIGURE 34   Two-dimensional projection of sheet silicate polytetrahedra as viewed from above, with atoms drawn to correct relative size. Silicon atoms are concealed beneath unshared oxygen atoms.

OH

unshared O

shared O

ITQ 18   If you examine Figure 35a, you will see that there are *six* magnesium atoms for every *eight* silicon atoms in the adjacent sheets. Can you suggest a chemical formula for this mineral?

Talc, or soapstone (Figure 35a), has a structure in which pairs of sheets are bonded together by magnesium ions to form what is known as the 'talc sandwich'. Talc is one of the softest minerals known (which is why it is used in talcum powder) and it has a greasy feel due to the ease with which the sandwiches slip past each other along the planes indicated with a heavy arrow in Figure 35a. As you can see, there are no ionic bonds to hold the sandwiches in position and the very weak physical forces between the sandwiches are easily overcome.

talc

Although talc is not one of the major silicate minerals, it provides a stepping stone to understanding the important group known as *micas*. In micas, *one* in every *four* of the sheet-forming silicon atoms is replaced by aluminium. (Remember that we introduced the subject of aluminium substitution in tetrahedral sites in ITQ 13.)

micas

What will be the general formula for the anion complex of mica sheets?

If we replace one in every four silicon atoms by aluminium, *the general formula for the complex becomes* $[(AlSi_3O_{10})(OH)_2]^{7-}$. Notice that this introduces another negative charge, so if we develop a talc-like mineral, there will still be a net negative charge. To balance this, one potassium ion enters the structure. This gives us the formula $K(Mg, Fe)_3AlSi_3O_{10}(OH)_2$.

We have put (Mg, Fe) in parentheses to indicate that there is a solid solution series between 100 per cent Mg and 100 per cent Fe ions in the octahedral sites. Members of this series are known as *biotite* micas: they are usually dark brown in colour and, like all micas, are *sheet-like* (*CB*, Plate 17).

biotite

56

Can you see from Figures 35a and 35b that the magnesium ions are in octahedral (6-fold) co-ordination with oxygen and $(OH)^-$ groups? What happens to the formula above if, instead of magnesium atoms, aluminium is incorporated into octahedral co-ordination sites?

Aluminium atoms are trivalent, so there can only be two of them for every three magnesium atoms in Figure 35b. In other words, one in every three octahedral sites will be left vacant. This gives us the chemical formula of white mica, or *muscovite*, $KAl_2AlSi_3O_{10}(OH)_2$, in which, for clarity, we have kept the octahedrally and tetrahedrally co-ordinated aluminium atoms separate.

muscovite

Now look at the position of the potassium ions in the mica structure (Figure 35b). Potassium forms a very large ion (radius = 133 pm) and occurs in 12-fold co-ordination in micas. (The atoms of potassium in Figure 35b are recessed into the paper and only half of the co-ordination linkages are shown.) The effect of KAl substitution for Si upon the physical properties of mica, as compared with talc, should now be plain. In mica, the layers or double sheets will not slide against each other because the potassium ions are holding them together. If you try to fold or bend a sheet of mica, it will spring back or, if bent far enough, will snap. All the same, the ionic bonding between one layer and the next is very weak, since each potassium is only singly charged and has 12-fold co-ordination. So there is a very well-developed *layer cleavage* in all micas (along the large arrow in Figure 35b) and the mineral can be cleaved to give transparent sheets (*CB*, Plate 17) of less than 0.01 mm thickness. So the crystal chemistry of mica elegantly explains its physical properties. These are exploited in another way when mica is used as an electrical or a thermal insulator; it is very difficult for heat or electricity to be conducted across the strong cleavage planes.

Would you expect micas to have low or high densities?

The very loosely compacted, open structure of sheet silicates gives them relatively *low* densities ($2\,500$–$3\,300\,kg\,m^{-3}$) and only iron-rich biotites fall in the upper end of this range. As you can also see from the chemical formulae in Table 8, the ratio of cations to silicon atoms in sheet silicates is the same as or a little lower than in chain silicates—we shall come back to this in Section 3.3.5.

Also under the heading of sheet silicates, there is another important group that we should mention: the *clay minerals*. For reasons that we are not going into here, clay minerals do not generally form crystals large enough to be seen without a microscope; they occur naturally as fine-grained aggregates (powders), coloured white, brown or grey depending on composition. In freshwater they form colloidal suspensions of very fine particles (c. 1–2μm in size), but in seawater, which is a strong electrolyte (a strongly ionized solution), they flocculate (coagulate) and are precipitated as muds. Flocculation takes place because the clay particles often have unsatisfied surface electrical charges which cause them to accumulate in the presence of the electrolyte. After precipitation they may become buried and the water may be gradually squeezed from within and between the particles until *clays* or *mudstones*[A] result. You should realize, therefore, that clays are involved in surface geological cycles: in fact, they are produced by the breakdown of the *primary* rock-forming silicate minerals which we have been considering so far. The processes of mineral (and rock) formation are considered in more detail later—in Section 4 of this Block and, in more detail, in later Blocks—here we introduce two simple structures, those of the clay minerals *montmorillonite* and *kaolinite**.

clay minerals

montmorillonite   **kaolinite**

The structures of talc and montmorillonite (Figure 36a) are almost identical, except that the former has magnesium atoms in the sandwich whereas montmorillonite has aluminium atoms in octahedral co-ordination. The chemical formula of montmorillonite, therefore, is $Al_2(Si_4O_{10})(OH)_2 \cdot nH_2O$. The nH_2O group added at the end indicates that montmorillonite, like many clay minerals, can take in a variable amount of water between its double layers and this makes the structure *swell*, resulting in a very low-density mass. The structure can also accommodate loosely held (not bonded) cations, such as Ca^{2+}, K^+, Mg^{2+}, Na^+, in the same structural positions as water molecules, and this often results in a very complex chemistry.

---

* Strictly, kaolinites are a group of clay minerals whereas montmorillonite is one of the smectite group of clay minerals. These points will be developed in Block 5.

57

The structure of *kaolinite* (Figure 36b) differs from that of montmorillonite (and talc) in comprising only *single sheets of tetrahedra* rather than sandwiches. However, instead of a second silicate layer above the aluminium atoms (as in montmorillonite) there is a layer of hydroxyl groups: so kaolinite contains only structurally bonded water (compare montmorillonite which contains both bonded and un-bonded water). You should satisfy yourself, from Figure 36b, that the chemical formula of kaolinite is $Al_4Si_4O_{10}(OH)_8$.

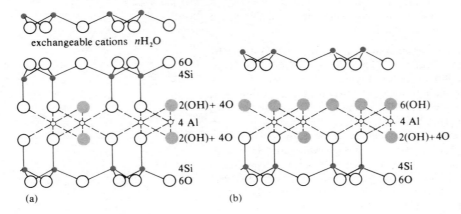

(a)   (b)

FIGURE 36 Two-dimensional projections of cross-sections through the structures of two clay minerals: (a) montmorillonite; (b) kaolinite. The rows of atoms are labelled and co-ordination bonds around aluminium atoms are indicated by dashed lines.

Finally, we would like to draw your attention to the fact that the double chain (amphibole) and sheet silicates are the two main groups of rock-forming minerals that are hydrous; *that is, they contain structurally bonded water.*

hydrous minerals

3.3.4 Framework silicates—quartz and feldspars

In *framework silicates*, all four of the vertices of each tetrahedron are shared, to form three-dimensional framework structures. After your attempts to make models of all the structures so far considered, from separate tetrahedra to sheet-like polytetrahedra, you might wonder how it is possible to make up a complete framework. We advise you not to try—but to look at *CB*, Plate 18, where we have done it for you. There is one important feature you must take note of: *every oxygen ion is now bonded to two silicon ions*, and, of course, every silicon ion is bonded to four oxygen ions. Therefore, the general formula for framework silicates must be SiO_2.

framework silicates

You may recognize this as the chemical formula of the mineral *quartz* which most commonly forms clear, glassy crystals (*CB*, Plate 19) which are very hard and have no cleavage. This is because the network of Si—O bonds in the quartz structure is uniformly strong in three dimensions.

quartz

There are several other minerals that have identical chemical compositions to quartz; these are the polymorphs of SiO_2. We mentioned a few in Section 3.2.3, such as the high-pressure form, *stishovite*, and the high-temperature form, *tridymite*. Collectively, these are known as the *silica minerals* and they include more exotic forms such as the hydrated silica mineral *agate* (*CB*, Plate 20): $SiO_2 \cdot nH_2O$.

silica minerals

The densities of all the silica minerals are low (c. $2\,650\,kg\,m^{-3}$). There are two reasons for this: in comparison with olivine, for example, quartz has a slightly less closely packed structure and, most important, it contains no elements (such as iron) with large relative atomic masses.

ITQ 19 Silica has a three-dimensional framework structure that is electrically neutral. What happens if the silicon in one tetrahedron in four is replaced by aluminium?

Feldspars (*CB*, Plate 21) are the most important constituents of many rocks found at the Earth's surface. In the *alkali feldspars* Al^{3+} replaces Si^{4+} in one out of every four tetrahedra, and the charges are balanced by the presence of one Na^+ or one K^+ per four tetrahedra, giving the following minerals:

alkali feldspars

orthoclase	$KAlSi_3O_8$
albite	$NaAlSi_3O_8$

You know from ITQ 14(b) that ionic substitution of K^+ by Na^+ is *very limited* because of the large difference in size of these cations. This means that, in rocks, the two minerals form *separately* (although, at high temperatures, around 800 °C, more extensive ionic substitution does occur).

However, there is a third 'end-member' feldspar that we need to consider: $CaAl_2Si_2O_8$, called *anorthite.*

> Will there be a solid solution series between anorthite and either of orthoclase and albite?

The similarity of ionic radii for Ca^{2+} and Na^+ (99 and 97 pm, respectively, from Figure 21) allows complete ionic substitution to occur between albite and anorthite to form the solid solution series known as the *plagioclase feldspars*. Notice one other point: as Na^+ is replaced by Ca^{2+}, so a second Si^{4+} atom (in every four) is replaced by Al^{3+}: this is the *coupled substitution* we discussed in Section 3.2.2. You will have realized by now that the two feldspar series have one component in common: sodium feldspar, $NaAlSi_3O_8$. There are very few natural occurrences of this *pure* end member in the plagioclase series because there is extensive solid solution behaviour. Both the range in compositions of the feldspars as a whole and the range in their solid solution can be illustrated by a new kind of diagram: a *triangular composition diagram* (Figure 37). You will meet several more examples of such triangular graphs later in the Course.

plagioclase feldspars

triangular composition diagram

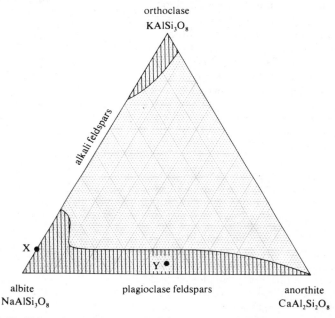

FIGURE 37 Triangular diagram of feldspar compositions (see the text for description of use). Most naturally occurring feldspars have compositions in the hatched areas; the points X and Y are described in the text.

This is how the feldspar diagram (Figure 37) works. At the three corners we have plotted the three 'end-member' compositions such that a feldspar consisting entirely of $KAlSi_3O_8$ would plot at the top apex of the diagram. An alkali feldspar containing 10 per cent orthoclase and 90 per cent albite would plot at the point marked X, which is nine-tenths of the way from $KAlSi_3O_8$ to $NaAlSi_3O_8$. Another feldspar, plagioclase this time, with 5 per cent orthoclase and 47.5 per cent of each of the two plagioclase end-members, would plot at the point Y. This is equidistant from the $NaAlSi_3O_8$ and $Ca_2Al_2Si_2O_8$ points, and nineteen-twentieths of the distance from $KAlSi_3O_8$ to the base line of the triangle. To plot these points accurately, triangular graph paper is normally used. You will be using these diagrams in Block 3.

The hatched areas in Figure 37 indicate the range in composition that occurs in feldspars that have been collected and analysed chemically. This illustrates and amplifies the conclusions made earlier that, in the alkali feldspars, ionic substitution of K^+ by Na^+ only occurs near the top of the diagram and replacement of Na^+ by K^+ only occurs near the albite corner. But, in plagioclases, there is complete ionic substitution to give all possible Na^+/Ca^{2+} ratios and, therefore, the solid solution series.

We have purposely said very little about the internal structures of feldspars because they become very complicated in detail. This happens particularly when there are large potassium atoms in the structure. The structure of all feldspars approximates to that of quartz (*CB*, Plate 18), the chief difference being that Al^{3+}, alkalis and/or calcium are also accommodated. This introduces bonds that are weaker than Si—O in quartz and, so, feldspars do have *cleavage*—usually in two directions nearly at right angles—but this is not nearly so well developed as in many chain and sheet silicates.

Finally, we should note that feldspars have densities ($2\,600–2\,800\,kg\,m^{-3}$) that are very similar to that of quartz; this is not surprising in view of their structural and chemical similarities to quartz. *Collectively, the light-coloured, low-density minerals that are rich in silicon and aluminium, may be termed* salic minerals (*as you will find in Table 9*).

salic minerals

3.3.5 The chemical compositions of silicates

Now that we have surveyed the individual structures and cation contents of the silicate minerals, we shall look at their chemical compositions—in terms of oxides, like the data in Table 3. Thinking of composition in terms of oxides is particularly useful when, in Section 4 and later Blocks, we come to describe the relationships between mineralogical and chemical compositions during the evolution of, for example, silicate liquids or *magmas*[A]. Most of the minerals we have described may form by crystal nucleation and growth from liquid magmas and the types of minerals that grow in a particular liquid are controlled by the concentrations of each free atomic species in that liquid. Thus, for example, if there is a high concentration ratio of magnesium atoms to silicon atoms, olivine crystals nucleate; at lower values of this ratio, pyroxenes may form preferentially (see ITQ 17) and, of course, liquids that are very rich in silicon may form quartz. As you will realize, most igneous rocks (rocks formed by cooling of magma) contain mixtures of crystals that formed either together, or successively, as the chemical composition of the melt changed. (This will be discussed in more detail in Section 4 and Block 3.) This rather simple approach to thinking about crystal formation in liquids leads to the conclusion that *the chemical compositions of igneous rocks, and the minerals they contain, must reflect the composition of the liquid from which the minerals crystallized.* So it is important that you have some knowledge of the chemical compositions of minerals.

Now look at Table 9 and compare the chemical analyses given there with the formulae for the different minerals summarized in Table 8. A useful generalization is that the combined proportion of silica SiO_2 and alumina Al_2O_3 (which partially replaces SiO_2 in amphiboles, micas and feldspars) generally increases down Table 9 at the expense of other chemical elements. You found this relationship for olivines, pyroxenes and amphiboles in ITQ 17. Notice that, in Table 9, concentrations are quoted in weight per cent. For example, although the ratio of FeO + MgO : SiO_2 in olivines is 2:1 by numbers of molecules, the ratio is less than 2:1 by mass because

TABLE 9 Approximate chemical compositions of some common silicate minerals (weight per cent)

Minerals	SiO_2	Al_2O_3	FeO + MgO	CaO	$Na_2O + K_2O$	H_2O
FERRO-MAGNESIAN						
Mg-rich olivine	40	—	60	—	—	—
Pyroxene	50	—	35	15	—	—
Amphibole	40	10	30	15	3	2
Biotite mica	36	20	30	—	10	4
SALIC						
Calcium-rich plagioclase	54	29	—	12	5	—
Sodium feldspar (albite)	68	20	—	—	12	—
Potassium feldspar (orthoclase)	65	18	—	—	17	—
Muscovite mica	47	37	—	—	12	4
Quartz	100	—	—	—	—	—

MgO (molar mass* = 40) is lighter than SiO_2 (molar mass = 60). In pyroxenes, the ratio of silicon atoms to cation atoms is 1:1 and this is reflected in the chemical analysis quoted in Table 9. Similar comparisons between formulae and chemical composition can be made for all the other minerals. *The analyses given in Table 9 are averages* and should not be interpreted too rigidly, especially in cases where solid solution behaviour leads to a wide range of possible chemical compositions.

Now try out your understanding of the chemical compositions of minerals with the following ITQ.

ITQ 20 Can you account for the following two features in Table 9:

(a) the increase in Al_2O_3/SiO_2 from sodium to calcium plagioclase?

(b) the decrease in SiO_2 *and* Al_2O_3 from sodium to potassium feldspar?

(*Hint*   The relative atomic mass of potassium is 39; that of sodium is 23.)

Before we leave this Section, it is worth spending a moment looking back to the probable composition of the whole Earth (Figure 13, p. 31) and asking which silicate minerals are likely to be common in the Earth. To make a simple estimate, bear in mind that some of the Earth's iron is in the silicate layer (Figure 8 and Section 3.1). Say this amounts to a quarter of the 25 per cent iron by mass in the whole Earth (Figure 13). This gives us the following relative masses of the three main 'cations'; each mass is divided by the atomic mass to obtain the numbers of atoms:

	Si	Mg	Fe
Relative masses	14.5	12.5	6.25
Relative numbers of atoms	0.52	0.52	0.11

On this basis, the ratio of Si: Mg + Fe atoms in the silicate layer of the Earth is 0.52:0.63 or 1:1.2. From your answer to ITQ 17, which minerals are likely to be most common?

A mixture of pyroxenes and olivines might seem to fit this prescription because the Si:Mg + Fe ratio is 1:1 in the former and 1:2 in the latter. By the time some of the 'other' elements in Figure 13 have been taken into account, the ratio of Si:other cations becomes a little closer to 1:2. As you will find in Block 2, *olivine is by far the most abundant mineral in the Earth's mantle, followed by pyroxene and a little feldspar. In the crust, however, the ratio of silicon to other cations is greater than 1:1, causing other minerals, particularly feldspar and quartz, to be most important.*

3.3.6 Bond strengths in silicate minerals

In Sections 3.3.1–3.3.4 we made passing reference to relative bond strengths in silicate minerals. In Section 4 you will find that a quantitative scale of cation–oxygen bond strengths is extremely useful in determining the fate of these minerals in water-affected *surface* environments on the Earth. This is because the familiar processes of chemical and mechanical *weathering* break down pre-existing rocks and the minerals they contain to form a mixture of *soluble materials*—cations liberated from primary minerals—and *insoluble particles*—usually mineral fragments that resist chemical breakdown. The stronger the bonds in a mineral, the more resistant it is to chemical weathering.

What factors are likely to influence the strengths of cation–oxygen bonds in minerals?

You probably realize that the size of the electrical *charge* on a cation will influence bond strength: the greater the charge, the stronger the bond. But each cation 'shares' its charge between the oxygen ions that surround it; that is, those to which it is co-ordinated.

Would you expect bond strength to go up or down with co-ordination number?

---

*The term 'molecular weight', which you may meet outside this Course, means the same as 'molar mass'.

The more bonds there are between oxygen ions and a particular cation, the smaller will be the strength of each bond. Therefore, *bond strengths decrease with increasing co-ordination number.* For our purposes, it is convenient to use a relative scale of bond strengths where:

$$\text{relative bond strength} = \frac{\text{charge}}{\text{co-ordination number}}$$

relative bond strength

ITQ 21   Table 10 lists some of the major cations found in silicate minerals. We would like you to complete the table by using the most usual co-ordination site (quoted in the Table) in silicate minerals for each cation to identify the relative bond strength for each cation compared with silicon in tetrahedral sites. (Note that we have identified two co-ordination states for aluminium.)

TABLE 10   Common cations, their co-ordination states (relative to oxygen) and bond strength (relative to silicon–oxygen bonds) in silicate minerals (for use with ITQ 21)

Cation	Co-ordination state	Relative bond strength
Si^{4+}	Tetrahedral (4-fold)	
Al^{3+}	Tetrahedral (4-fold)	
Al^{3+}	Octahedral (6-fold)	
Mg^{2+}	Octahedral (6-fold)	
Fe^{2+}	Octahedral (6-fold)	
Ca^{2+}	Cubic (8-fold)	
Na^{+}	Cubic (8-fold)	
K^{+}	12-fold	

Make sure that you have Table 10 correctly filled in before reading any further.

Table 10 shows that the first bonds to break in a mixture of silicate minerals will be K—O bonds, which occur mainly in micas and feldspars. These will be followed by other bonds in feldspars (Na—O and Ca—O) and then bonds involving ferromagnesian minerals—chain-linking cations in pyroxenes and amphiboles (Figures 30 and 32)—and sandwich-linking cations in sheet silicates (Figure 35). Note, however, that Si—O and Al—O bonds are strong, so that chains, sheets and frameworks themselves are not broken down easily. *This is why the degradation of primary polytetrahedra leads to the release of certain cations in solution but also leaves smaller insoluble particles made of mainly silicon, aluminium and oxygen*—mainly clay minerals and quartz. These will be discussed in more detail in Section 4.

But what about minerals based on single tetrahedra—the olivines? Look again at Figure 28. Once the cations linking individual tetrahedra are released, all that remain are minute fundamental silicate particles that are released as colloids, behaving as if they were in solution. We shall come back to this later, but you should notice that the most complex polytetrahedra are most resistant to *complete* chemical breakdown. Thus, of the common silicates in Table 8, *quartz is most resistant, followed by chain, sheet and other framework silicates; olivines are least resistant to chemical weathering.*

The main features of the silicate minerals that you have been studying in Section 3.3 are given in Tables 8, 9 and 10. These should be used as summaries for reference and revision purposes.

3.4 Non-silicate minerals

Although the silicate minerals are by far the most abundant in the crust and mantle, there are many other kinds of *non-silicate minerals*, some of which are important at the surface. These may occur in low concentrations in common rocks as, for example, small amounts of iron oxides in igneous rocks. Others may become concentrated into quite considerable deposits, as in the formation of sulphide ore deposits and evaporite salts or, more significantly, as in the accumulation of carbonate deposits, *limestones*[A], which will be discussed in more detail in Section 4.

Most non-silicates are classified according to the anion or anion complexes they contain (see Table 11). The crystal structures of non-silicates follow similar rules to those for silicates: they have polymorphs (for example, $CaCO_3$ can be found as calcite and aragonite; carbon as graphite and diamond; as discussed in Section 3.2.3) and solid solution series, and they obey familiar co-ordination and valency rules. However, there are few examples of giant polyanions like those that occur in silicates; most non-silicates have separate anions or simple anion complexes bonded by cations (as in olivine silicates). We have already discussed the structures of two non-silicates: graphite and diamond (Figure 25, p. 48). Here are a few more examples that you will meet again during this Course.

We will start with *fluorite* (CaF_2), in which the bonds are strongly ionic. The near-perfect cubic crystals of fluorite are shown in *CB*, Plate 22.

non-silicate minerals

fluorite

> **ITQ 22** Use the ionic radii of Ca^{2+} and F^- in Figure 21 to predict the co-ordination number of calcium in fluorite. Hence show that the co-ordination number determines the structure and morphology of fluorite crystals.

TABLE 11 Some non-silicate minerals. (You are not expected to remember this Table in detail.)

Mineral group	Typical examples
Single elements	C (graphite, diamond) S ⎫ Au ⎬ native elements Pt ⎭
Oxides	Fe_2O_3 (hematite) Fe_3O_4 (magnetite) TiO_2 (rutile) $FeTiO_3$ (ilmenite) SnO_2 (cassiterite) $FeCr_2O_4$ (chromite)
Hydroxides	AlO(OH) (boehmite) ⎫ both common in bauxite (Al_2O_3) $Al(OH)_3$ (gibbsite) ⎭
Sulphides	FeS_2 (pyrite) FeS (troilite) $CuFeS_2$ (chalcopyrite) ZnS (sphalerite) PbS (galena)
Sulphates	$BaSO_4$ (barytes) $SrSO_4$ (celestine) $CaSO_4 \cdot 2H_2O$ (gypsum) $CaSO_4$ (anhydrite)
Carbonates	$CaCO_3$ (calcite, aragonite) $CaMg(CO_3)_2$ (dolomite) $FeCO_3$ (siderite) $Cu_2CO_3(OH)_2$ (malachite)
Phosphates	$Ca_5(PO_4)_3(OH)$ (apatite) (La, Ce, Th)PO_4 (monazite)
Halides	CaF_2 (fluorite) NaCl (halite, rock salt)

Figure 38 shows that the cubic framework of fluorite is built from fluorine atoms and, in the *large* cube shown, there are *eight* available cubic co-ordination sites at the centres of the eight *small* cubes. But because there must be a $Ca^{2+}:F^-$ ratio of 1:2 in order to preserve electrical neutrality, only four of the eight sites are filled. One example each of co-ordination bonds around a fluorine and a calcium ion are shown.

Now, let us look at *calcite*. In the two calcium carbonate polymorphs, the carbonate anion complexes $(CO_3)^{2-}$ contain covalent carbon–oxygen bonds and are indicated in the calcite structure (Figure 39) as triangles. Calcium atoms are octahedrally co-ordinated by oxygen, and the stacking of the cations and anion complexes gives a rhombohedral shape to the structure (Figure 39) and to the crystals (*CB*, Plate 23). In aragonite, the structure is similar to that of calcite, but slightly more closely packed, giving a higher-density mineral that is stable at high pressure (Figure 54, p. 93).

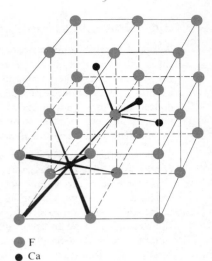

F
Ca

FIGURE 38   The fluorite structure (CaF_2).

calcite

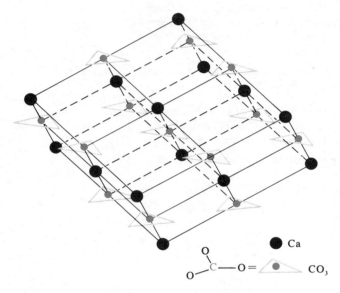

FIGURE 39   The calcite structure $(CaCO_3)$, containing complex $(CO_3)^{2-}$ anions bonded to calcium.

Like fluorite, *pyrite* (FeS_2) has a cubic structure (*CB*, Plate 24) but, in this case, there is an equal number of Fe^{2+} cations and S_2^{2-} *double anions*, which alternate in the structure (Figure 40). The reason why S_2^{2-} forms a double anion is not complicated.

> If we tell you that the separation of the two sulphur nuclei in each pair is about 214 pm, whereas the sum of the two ionic radii (for separate atoms) would be 370 pm, does this suggest a reason?

The two atoms are closer together than two similar sized ions because their outer electron orbitals are shared. Iron sulphide is a good example of *covalent* bonding in minerals (note the small electronegativity difference between iron and sulphur plotted in Figure 16 (p. 35)). Despite this electron sharing, the individual S_2^{2-} groups behave as anion complexes whose long axes are oriented as in Figure 40. (The groups are shown as 'dumbbells' but, of course, each really comprises two overlapping spheres.) Pyrite has a bright yellow metallic lustre and it is not without good cause that it is known as 'fool's gold'.

pyrite

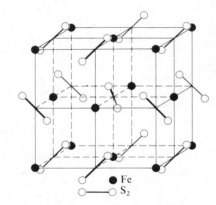

Fe
S_2

FIGURE 40   The pyrite structure (FeS_2), containing complex S_2^{2-} anions bonded to iron.

This brings us to the end of our review of minerals: the basic 'building blocks' of which the Earth is made. We are now poised to use all this information on the physics and chemistry of minerals to find out much more about the inside of the Earth. This is a central theme in this Course and in Section 4 we shall explore in more detail the relationship between the geochemistry of rocks and their mineral content. Section 4 will also start to show that the types of rock we find at the Earth's surface are determined by their mode of origin.

3.5 Objectives for Section 3

Now that you have completed Section 3, you should be able to:

1 Recognize valid definitions of the terms flagged in the margin of this Section.

8 Show how the geochemical classification of the elements is directly related to the fundamental Periodic, bonding and electronic characteristics of each element.

9 Give a fully reasoned account of why a particular chemical element is likely to be concentrated in the metallic, sulphide-rich, or silicate portions of the layered Earth.

10 Explain the importance of X-ray diffraction in determining the internal structures of minerals on an atomic scale, and carry out simple calculations involving Bragg's equation.

11 Describe, with reasons, the variations in ionic size of chemical elements and show how the radius ratio in ionic bonding determines the co-ordination states of cations and anions in crystal structures.

12 Distinguish, with examples, the factors that favour isomorphism, polymorphism, partial ionic substitution and solid solution series.

13 Build simple models of silicate polytetrahedra with the Home Experiment Kit and describe the structures of the complex silicate anions.

14 Explain, with examples, the relationship between atomic structural properties, physical features (e.g. cleavage and density) and cation content of the five main groups of silicate minerals (Table 8).

15 Give a reasoned account of the variations in chemical composition among silicate minerals.

16 Account for the variations in relative cation bond strengths in silicate minerals and use them to predict which elements will be most easily freed during chemical breakdown.

17 Explain how the principles used to study the main groups of silicate minerals may be applied to some non-silicate minerals.

Apart from Objective 1, to which they all relate, the sixteen ITQs in this Section test these Objectives as follows: ITQ 7: Objectives 6 and 9; ITQ 8: Objective 8; ITQ 9: Objectives 8 and 9; ITQ 10: Objective 10; ITQ 11: Objective 11; ITQ 12: Objective 11; ITQ 13: Objectives 11 and 12; ITQ 14: Objective 12; ITQ 15: Objectives 12 and 17; ITQ 16: Objective 13; ITQ 17: Objectives 14 and 15; ITQ 18: Objective 14; ITQ 19: Objective 14; ITQ 20: Objectives 12 and 15; ITQ 21: Objective 16; ITQ 22: Objective 17.

You should now do the following SAQs, which test other aspects of these Objectives.

SAQ 7 (*Objectives 8, 9 and 11*) The electronegativities, ionic radii and ionic charges of four unknown elements and iron are given below:

Element	Electronegativity	Ionic radius (pm)	Ionic charge
A	2.4	139	2+
B	0.9	107	1+
C	2.0	75	2+
D	1.4	46	4+
Fe	1.8	74	2+

(a) Place elements A–D in their most likely order of ease of oxidation relative to iron.

(b) What is the most likely co-ordination state for each element, assuming that each would be able to bond ionically with oxygen (radius 140 pm)?

(c) From (a) and (b), into which of the following categories is each element (A–D) likely to fall? (Bear in mind that iron is usually chalcophile; see Section 3.1.3.)

(i) mainly ionic bonding to *form* anion complexes with oxygen;

(ii) mainly ionic bonding to *link* anion complexes;

(iii) mainly covalent bonding with sulphur;

(iv) most likely to be simply metallic.

When a statistically significant number of points has been counted (usually c. 1 000), the results are converted into a modal analysis. It is assumed for this purpose that the area represented by each mineral in thin sections is proportional to its volume in the rock. Notice the very striking contrast between these two modal analyses: *granite is primarily composed of quartz and alkali feldspar, whereas basalt is made mainly of pyroxene and plagioclase feldspar* (like its coarse-grained equivalent in Table 12, gabbro[A]).

It will be obvious that in any rock analysis, each mineral present (for example, those in Table 13a) contributes various amounts of different chemical elements to the bulk chemical composition of the rock (Table 13b). If you compare the appropriate chemical analyses for the common silicate minerals in Table 9 with the two modal analyses in Table 13a, you can begin to account for the major features in the chemical analyses of the two rocks. Try that out in ITQ 24.

> **ITQ 24** (a) Look at the minerals listed in Table 13a. Why is SiO_2 so much more abundant in granite than in basalt?
>
> (b) Which minerals in each rock provide the iron and magnesium in the analyses of Table 13b? From your answer, can you explain why basalt contains more Fe and Mg than granite?
>
> (c) Why does basalt contain so much more calcium than granite?

Because SiO_2 was formerly regarded as an acidic oxide, the rocks in which it is most abundant, such as granites, were defined as *acid*. Although silica is structurally bonded in silicates and seldom forms *aqueous* solutions that are distinctly acidic, this terminology is still in common use. Rocks such as basalt, in which SiO_2 was thought to be 'balanced' by weak basic oxides like CaO and MgO, were termed *basic*, whereas rocks with high MgO contents and the lowest SiO_2 contents (c. 40 per cent), such as peridotite, were called *ultrabasic*. These terms should not be confused with conventional chemical use of the term 'acidity' which describes the concentration of hydrogen ions in aqueous solution (pH[A]), but they *do* relate to the amount of silica in a rock analysis, according to the following convention *which is applied only to igneous rocks, and particularly their magmas*: **acid rocks**

basic rocks
ultrabasic rocks

SiO_2 over 65 per cent	acid
52–65 per cent	intermediate
45–52 per cent	basic
below 45 per cent	ultrabasic

intermediate rocks

The close relationship between chemical and mineralogical compositions that you examined in ITQ 24 means that, for example, an acid magma will crystallize salic minerals and may form the rock granite. Conversely basic magmas will crystallize ferromagnesian minerals and may form the rock basalt (or gabbro).

Table 14 gives a more complete list of the chemical compositions of coarse-grained, intrusive igneous rocks.

> Which rocks in Table 14 are acid, which intermediate, which basic and which ultrabasic?

TABLE 14 Approximate chemical compositions (weight per cent) for some common igneous rocks (coarse-grained varieties only)

	Granite	Granodiorite	Diorite	Gabbro	Peridotite
SiO_2	72.1	66.9	53.9	48.4	43.5
TiO_2	0.37	0.57	1.5	1.3	0.81
Al_2O_3	13.9	15.7	15.9	16.8	4.0
Fe_2O_3	0.86	1.3	2.7	2.6	2.5
FeO	1.7	2.6	6.5	7.9	9.8
MnO	0.06	0.07	0.18	0.18	0.21
MgO	0.52	1.6	5.7	8.1	34.0
CaO	1.3	3.6	7.9	11.1	3.5
Na_2O	3.1	3.8	3.4	2.3	0.56
K_2O	5.5	3.1	1.3	0.56	0.25
P_2O_5	0.18	0.21	0.35	0.24	0.05
H_2O	0.53	0.65	0.80	0.64	0.76

You should have recognized the granite and granodiorite as acid, the diorite as intermediate, the gabbro as basic and the peridotite as ultrabasic.

Now look at Figure 41, which summarizes the mineralogy, grain size and silica content of the main igneous rock types. Notice that igneous rocks characterized by *low silica contents* (*c.* 40 per cent) contain a high proportion of *ferromagnesian* minerals (shaded in Figure 41) whereas those with *high silica contents* (*c.* 70 per cent) contain mainly *salic* minerals. These observations really reiterate the conclusions you have already made in ITQ 24 about granite and basalt (gabbro), but Table 14 and Figure 41 fill in the gaps and show the *smooth progression of changing mineralogical and chemical compositions from acid to ultrabasic igneous rocks*. As you can see from Figure 41, the trends from granite to gabbro among coarse-grained rocks are similar, in mineralogical and chemical terms, to those among the fine-grained rocks from rhyolite (acid) to *andesite*[A] (intermediate) to basalt (basic).

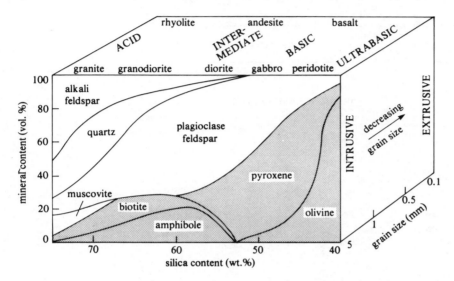

FIGURE 41   Block diagram showing the approximate mineral content, silica content and grain size of different igneous rocks. Shading indicates the volume of each rock type occupied by ferromagnesian minerals. (Note that the minerals present on the front face apply to all grain sizes.)

ITQ 25   (a)   Look again at Table 14 and note (from Figure 41) that, at the ultrabasic end of the range, peridotite is made almost entirely of olivine, but with some pyroxene and few salic minerals. Moving from right to left across Table 14 from peridotite towards granite, can you identify oxides that:

(i)   increase in concentration?

(ii)   decrease in concentration?

(iii)   increase and then decrease?

(b)   In each case (i–iii), try to explain why this is so in terms of minerals involved. (The chemical compositions of minerals are given in Table 9.)

By now, we hope you will be convinced that *the mineralogical compositions of igneous rocks are a function of their major element chemical constituents and vice versa*. Earlier, we noted also that the chemical compositions of *magmas* and the igneous rocks they produce by crystallization are closely related; this implied that the *crystals produced are determined by the chemical compositions of the magmas* (as you learnt in Section 3.3.5).

But it is now well known, from experimental work, that different minerals crystallize at different temperatures. What does this imply about the *crystallization temperatures of magmas*?

Table 15 gives some very approximate crystallization temperatures for the silicate minerals that are abundant in igneous rocks. You will notice that, for the olivine and plagioclase solid solution series, there is a *range* of crystallization temperatures: Mg-rich olivines and Ca-rich plagioclases crystallize at higher temperatures than their Fe-rich and Na-rich counterparts. In fact, as you may have noted from

Section 3.3, analogous solid solution series can be identified for pyroxenes, amphiboles and biotite as well, but they are less easily defined, and for simplicity we have omitted them from Table 15. But in general, Mg-rich ferromagnesian minerals in a particular group crystallize at higher temperatures than their Fe-rich counterparts. Check that you understand this by doing ITQ 26.

ITQ 26 (a) From Table 15, in what order would you expect the silicate minerals present in a granite and in a basalt to have crystallized? (See Table 13a for the appropriate minerals.)

(b) What does this imply about the approximate temperature ranges over which granite and basalt magmas crystallize?

TABLE 15 The sequences of crystallization of the common silicate minerals as a function of temperature

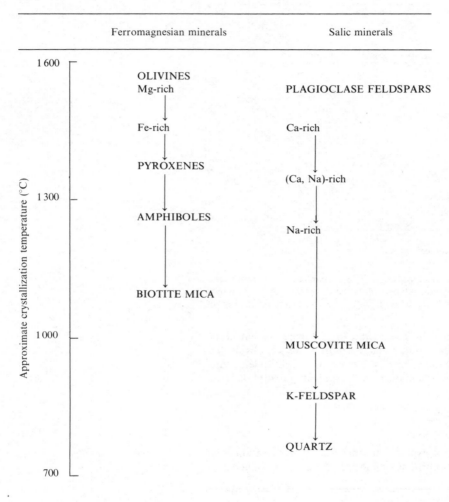

The temperature ranges that you deduced from Table 15 in ITQ 26 are actually a few hundred degrees centigrade higher than they are in nature: the normal crystallization temperatures of basic magmas are in the range 1 200–900 °C, and of acidic magmas are in the range 900–700°C. The temperatures in Table 15 were determined experimentally on specimens of pure minerals, whereas the magmas from which igneous rocks crystallize are complex mixtures of many minerals (see, for example, Table 13a). This has the effect of lowering the temperature at which each mineral crystallizes, although the *order* in which they crystallize is the same as in Table 15. The reasons for this behaviour are described and explained in detail in Block 3.

Until now, we have implied that the chemical compositions of magmas and the rocks they produce are virtually identical and, in many cases, this is quite true. But Table 15 shows that minerals crystallize from a particular magma in a well-defined sequence. We will look briefly at what happens to the *composition* of a magma when minerals begin to crystallize from it. Olivine contains 40 per cent SiO_2 and 60 per cent (FeO + MgO) according to Table 9, but basic magma may contain about 45 per cent SiO_2 and about 16 per cent (FeO + MgO) according to Table 13. Table 15 shows that olivine is the first mineral to crystallize from basic magma.

What do you think will happen to the content of SiO_2 in the remaining liquid on the one hand, and of (FeO + MgO) in the liquid on the other, as olivine crystals form and *separate* from it?

Because olivine crystals contain *less* SiO_2 than the liquid, the content of SiO_2 in the liquid will *increase* as olivine crystals separate from it. Conversely, because olivine contains *more* (FeO + MgO) than the liquid, the content of these constituents in the liquid will *decrease*. Now, if the olivine crystals were to be completely removed from the whole body of magma, say by sinking to the bottom on account of their high density, then the remaining liquid would be left enriched in SiO_2 and relatively depleted in (FeO + MgO). If this modified liquid were to migrate to some other part of the crust and to crystallize there, the resulting rock would not have the composition of the original basic magma because some of its constituents would have already been removed.

It is possible for an originally homogeneous basic magma to be modified in composition by the removal of successive 'crops' of crystals, or crystal fractions, to produce a whole range of rock types. This process of *fractional crystallization* or crystal fractionation can take place within a large mass of slowly cooling magma by the gravitational settling (literally, sinking) of crystals as they form. In fact, some enclosed bodies of crystallizing basic magma are known to have produced intrusions with layers ranging between the extremes of olivine crystals at the bottom and granite at the top of the intrusion. We shall be discussing the process of crystal fractionation and its influence on the major and trace element compositions of melts in much more detail in Block 3.

fractional crystallization

So far we have touched on the compositions of magmas and their relationships with the rocks they produce. But you may have been wondering where and how these magmas originate. The production of magma requires some kind of *melting* process—*the reverse of crystallization*. The geographical distribution of large bodies of igneous rock at the Earth's surface provides some clues about where to focus our attention on melting processes. The two most abundant igneous rock types at the Earth's surface are *basalt* and *granodiorite*. (Table 14 gives chemical analyses; remember that basalt and gabbro share the same composition.) In the simplest *plate-tectonic*[A] terms, basalt and granodiorite characterize *different* plate boundaries—the differences in their locations and their compositions provide good evidence that their parental magmas are fundamentally different.

Can you remember where oceanic plates are created and consumed?

The plates (consisting of both crust and uppermost mantle) are formed at ocean ridges, and they spread away from the ridges to active continental margins or island arcs where they are consumed. The two types of boundary are, respectively, *constructive*[A] and *destructive*[A].

Figure 42 shows that *basic magmas characterize constructive margins*, where basalt volcanoes form ocean ridges and are underlain by gabbro intrusions. *Intermediate to acid magmas characterize destructive margins*, where andesite volcanoes are underlain by granodiorite intrusions.

Here is a brief summary of the *processes* thought to be responsible for magma generation in these two locations; the details will be enlarged upon in later Blocks. It is thought that the mantle beneath ocean ridges undergoes partial melting due to the high temperatures associated with upwelling mantle convection cells (these will be discussed in more detail in Block 4). The material of the mantle is a form of peridotite which, after 10–20 per cent melting, yields *basic* magmas. They rise towards the surface where they either crystallize within the crust to form gabbro intrusions, or are erupted to form basalt lavas. Beneath the gabbro/basalt layer that makes up most of the ocean crust, there is a thick layer of unmolten peridotite (Figure 42). These two layers (basaltic ocean crust and (solid) upper mantle peridotite) move together as a rigid plate—the *oceanic lithosphere*[A]—until they reach a continental margin or island arc where they are *subducted* (resorbed) back into the mantle.

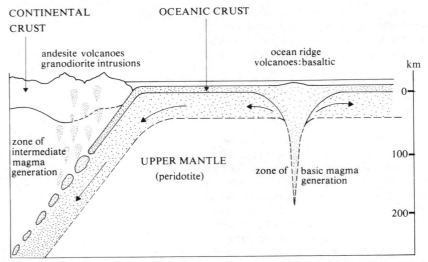

CONTINENTAL CRUST

OCEANIC CRUST

andesite volcanoes
granodiorite intrusions

ocean ridge
volcanoes:basaltic

km

0

zone of
intermediate
magma
generation

UPPER MANTLE
(peridotite)

zone of basic magma
generation

100

200

FIGURE 42   Schematic illustration of the oceanic plate cycle indicating the generation of basic oceanic crust at a spreading ridge by partial melting in the mantle, and a second zone of intermediate magma generation where the same crust is subducted back into the mantle beneath island arcs or active continental margins. Grey shading indicates oceanic lithosphere, which is made up of crust and uppermost mantle.

What do you think will happen to the basic rocks of the ocean crust as they are subducted?

First they are metamorphosed because they are subjected to increasing temperature and pressure, and then, with further subduction, the oceanic crust either *partially melts* or *releases a watery fluid*. The melt or fluid rises into the overlying mantle wedge (Figure 42), which may itself *partially* melt, and the net result is that *intermediate magmas* are generated. These rise into the continental crust where they may be erupted, forming *andesite* volcanoes; or they may be intruded as coarsely crystalline intrusions that range in rock type and composition from *diorite to granite* (Table 14). So the source of intermediate and acid magmas lies mainly within or above the subducted slab of oceanic lithosphere but there is good evidence that, sometimes, the continental crust may also partially melt in such locations. Notice that the generation of intermediate and acid magmas by partial melting within and above the subduction zone, produces melts richer in alkali elements and silica (granite and granodiorite—Table 14) compared with the source (gabbro and peridotite—Table 14). *Therefore, the result of partial melting is, in some ways, like fractional crystallization in reverse.* Minerals that crystallize at high temperatures (olivines, then pyroxenes and Ca-rich feldspars—Table 15) are just the ones that stay unmolten during partial melting. Can you now see that basaltic ocean crust is a transient feature of the Earth's surface? It is created and destroyed on a much shorter time-scale than the relatively permanent continental crust, but it serves as an intermediate step in the establishment of the continental crust at the expense of the mantle.

partial melting

The continental crust may, therefore, be thought of as the end-product of a chemical selection process in which the most easily melted components of the Earth's mantle are segregated into low-density continents (see Figure 8). The stages in this process are (i) the production of basaltic ocean crust from the mantle at spreading ridges, and (ii) the production of intermediate and acid magmas at destructive margins. This thumb-nail sketch of magma types and their products at the Earth's surface will be amplified later in the Course.

4.2 Sedimentary processes and products

Most sedimentary processes occur at the Earth's surface and are dominated by reactions involving *water*, usually in *oxidizing environments*. Sedimentary rocks result from a series of processes involving (i) the breakdown of pre-existing rocks and minerals by *physical*[A] *and chemical weathering*[A]; (ii) the *erosion* (or removal) of the solid and dissolved weathering products; (iii) their *deposition* as loose sediments, followed by (iv) long-term changes (known as *diagenesis*) that convert sediments into sedimentary rocks.

diagenesis

Most sediments accumulate around the shore-lines and on the continental shelves of land masses undergoing erosion: some, particularly the finer-grained varieties, accumulate as an uppermost layer of the ocean crust in the deep ocean basins (you will learn more about this in Block 2).

Our first step is to examine the products of the reactions between silicate minerals and water. Most of these reactions (e.g. equations 12 and 13) take place spontaneously, are irreversible and *exothermic*. Remember that an exothermic reaction is one that releases energy, leading to products that are of *lower energy* than the reactants. In such cases the *enthalpy of reaction*^{A} (ΔH) *is negative, since it is defined as:*

$$\Delta H = H \text{ (products)} - H \text{ (reactants)} \qquad (11)$$

This is shown diagrammatically in Figure 43; note that an *endothermic reaction* requires an energy input, has a positive ΔH (to be consistent with Le Chatelier's

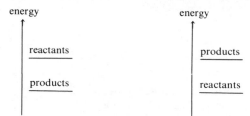

(a)   an exothermic reaction
H (products) $< H$ (reactants) so $\Delta H < 0$

(b)   an endothermic reaction
H (products) $> H$ (reactants) so $\Delta H > 0$

FIGURE 43 Schematic illustration of enthalpy changes at constant pressure and temperature for (a) an exothermic reaction and (b) an endothermic reaction.

principle), and leads to an *increase* in the energy of products. The evaporation of water is an endothermic process, whereas the condensation of steam into water is exothermic. This short digression into enthalpy of reaction should help you to understand that

(i)   *the breakdown of silicate minerals by exothermic weathering reactions readily takes place given suitable lengths of time when those minerals are exposed to water at the Earth's surface*;

(ii)   many of the reactions leading to the reconstruction of more complex minerals by metamorphism (which will be discussed in Section 4.3) are endothermic, and so require energy in the form of heat before they can take place. *During weathering, the chemical properties of the minerals involved determine whether they will go into solution, or give rise to solid products.* The relative bond strengths between their constituent elements (which were discussed in Section 3.3.6) are particularly important here. Look back at your completed Table 10 of relative bond strengths for cations in silicates (ITQ 21, p. 62).

Which one of the common rock-forming silicates would you expect to be *most resistant* to chemical breakdown?

We hope you didn't have too much difficulty in deducing that this will be *quartz*, because this mineral has only strong Si—O bonds.

So quartz will tend to be an *unaltered solid product* of chemical weathering.

But what will happen to feldspars during chemical weathering?

Feldspars do contain some weak bonds, those involving K^+, Na^+ and Ca^{2+} (Table 10). These elements will be released easily and will go into *aqueous solution*. This leaves the aluminium silicate parts of feldspar as solid components. (Look back at the answer to SAQ 12c, p. 97.)

These solid products of feldspar decomposition are the *clay minerals* (which you met in Section 3.3.3) and the breakdown of K-feldspar (orthoclase) can be illustrated by equation 12 (which you need not try to remember!):

$2KAlSi_3O_8 \quad + \quad H_2O + 2H_2CO_3$

orthoclase   weathering fluid
feldspar   (carbonic acid)

$\longrightarrow Al_2Si_2O_5(OH)_4 \quad + \quad 2K^+ + 2HCO_3^- + 4SiO_2(aq) \quad (12)$

kaolinite residue K^+ ions, HCO_3^- ions and silica in solution

The fact that silica is released means that some silicon–oxygen bonds must be broken when clay minerals form from feldspar. The same general features apply when the *mica* minerals are subject to chemical breakdown: *clay minerals and a soluble component are produced.*

Now consider the ferromagnesian minerals with less complex silicate structures than feldspars and micas: *olivines and pyroxenes.* What will happen to these during chemical weathering?

Although bonds between iron or magnesium atoms and oxygen are slightly stronger than the weakest bonds in feldspar, once they are broken, the remaining silicate material is in almost its most fundamental state—SiO_4 tetrahedra (in olivines) and SiO_3 chains (in pyroxenes), both of which are very small indeed. The total breakdown of olivine and pyroxene is therefore very rapid during chemical weathering and is reflected in their absence from sedimentary rocks (Table 18, answer to ITQ 23). We can express the breakdown of olivines as follows:

$$Mg_2SiO_4 + 4H_2CO_3 \longrightarrow 2Mg^{2+} + 4HCO_3^- + SiO_2(aq) + 2H_2O \qquad (13)$$

olivine   weathering fluid Mg^{2+} ions, HCO_3^- ions and silica in solution
(carbonic acid)

Similarly, the breakdown of the pyroxene chains is easily accomplished during chemical weathering. All the products are soluble and will be removed from the weathering site in solution, except for iron, which in oxidizing environments forms insoluble red–brown precipitates of iron hydroxide.

You should now appreciate that framework silicates are least easily broken down by the agency of chemical weathering whereas minerals with the smallest anion complex units are most easily decomposed. *Thus for silicate minerals the order of increasing stability during chemical weathering is remarkably similar to their order of crystallization from magmas* (Table 15). Silicate minerals that crystallize at high temperatures are unstable compared with those that crystallize at low temperatures.

ITQ 27   Given the same two igneous rocks as in ITQ 26, granite and basalt, which rock would you expect to decompose most rapidly as the result of chemical weathering? What will be the products in each case?

So the products of the chemical breakdown of primary silicate minerals are:

(i)   unaltered quartz grains;

(ii)   altered solid products, mainly clays but also including iron-rich precipitates;

(iii)   soluble species carried by river water: Ca^{2+}, Mg^{2+}, Na^+, K^+ and silica.

What will happen to each of these three types of weathering product during *erosion* (i.e. removal from the weathering site, usually by moving water)?

Clearly, the soluble (dissolved) weathering products are *separated* from the insoluble components during erosion—they are simply washed away. Also, the different solid components, mainly quartz and clay minerals, will be *sorted* in moving water by virtue of their contrasting *size and density.* Generally, quartz particles are much more coarse-grained than clay particles (which are $<2\,\mu m$), so quartz is transported by much more energetically moving water or else deposited. *Quartz-rich* sediments result from deposition in *quite high-energy environments*[A] (storm beaches, river terraces, etc.), whereas *clay-rich* deposits accumulate in almost still water, characteristic of low-energy environments, such as lagoons and ponds of flood water. The relationship between grain size and sediment type is shown in Figure 44 together with the rock types that result from diagenesis (compaction and cementation) of each sediment. Note that *silt* is a term meaning grain size intermediate between sand and clay and that *mud* is an informal term applying to a mixture of silt and clay (Figure 44).

The sorting of sedimentary particles by surface processes not only separates size fractions but also separates grains of different chemical compositions. This is because most coarse-grained (sand) deposits are quartz-rich and are therefore rich in SiO_2, whereas most fine-grained deposits (containing clay minerals) are complex mixtures of different elements. Table 16 contains major element chemical compositions for three typical, contrasting sedimentary rocks. Also shown, for comparison, is one estimate of the average composition of the continental crust, thought by many to be close to that of intermediate igneous rocks—as you can see for yourself by checking it against Table 14. Compared with this average crustal composition, many sedimentary rocks are, chemically, relatively pure. For example, a pure quartz sand would literally contain 100 per cent SiO_2, though many *sandstones*[A] are impure like the one in Table 16. The greywacke in your Home Experiment Kit (specimen 12) is still less pure; it contains some unaltered feldspar, illustrating an important point: *chemical breakdown takes place only if the right conditions persist for long enough.*

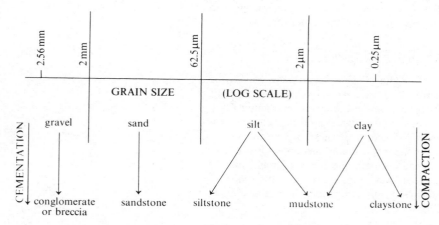

FIGURE 44 Grain-size variations among sediments and sedimentary rocks resulting from the solid products of chemical weathering (mainly coarser quartz and finer clay minerals).

The *mudstone*[A] analysis in Table 16 is chemically more complex, and there are several reasons for this. First, clay minerals have polyanions that contain both aluminium and silicon (Figure 36). Second, as you may recall from Section 3.3.3, clay minerals can accommodate cations such as Na^+, K^+, Ca^{2+} and Mg^{2+} into positions between the layers. Often they also collect such cations on their surfaces in a process known as *adsorption* (this is a process we shall discuss in Block 5). This means that *some of the soluble products of chemical weathering may be incorporated into clay-rich sediments*, which helps to explain why mudstones are chemically less 'pure' than sandstones and limestones (Table 16).

In fact, mudstones are often regarded as the geochemical 'rubbish tip' of surface processes. Clay minerals already consist of Al–Si–O complexes, and these pick up cations in the proportions in which they are available in surface waters. As those proportions depend on what is available to be dissolved in the weathering cycle, it is not surprising that mudstones take on the approximate chemical composition of the entire crust.

TABLE 16   Typical chemical compositions in weight per cent for some sedimentary rocks. Also shown for comparison is an estimate of the average composition of the continental crust.

	Average continental crust	Sandstone	Mudstone	Limestone
SiO_2	59.4	78.4	58.5	5.2
TiO_2	1.1	0.25	0.66	0.06
Al_2O_3	15.4	4.8	15.5	0.81
Fe_2O_3	3.1	1.1	4.1	0.54
FeO	3.8	0.30	2.5	0.00
MgO	3.5	1.2	2.5	7.9
CaO	5.1	5.5	3.1	42.6
Na_2O	3.9	0.45	1.3	0.05
K_2O	3.2	1.3	3.3	0.33
P_2O_5	0.30	0.08	0.17	0.04
H_2O	1.2	1.6	5.0	0.77
CO_2	0.10	5.0	2.7	41.6
C	0.00	0.00	0.81	0.00

So far we have concentrated almost entirely on the solid products of chemical weathering. What about the dissolved material?

Look again at Table 16. Can you see where most of the calcium released by chemical weathering is redeposited?

The answer is quite clear: it occurs in *limestones*[A] as $CaCO_3$, which is represented in Table 16 by two components, CaO and CO_2.

But how does this soluble calcium become incorporated into limestones?

There are two major ways in which all dissolved species in ocean water are removed (or else the ocean water would rapidly become more saline) and these are:

(i)　by *direct chemical precipitation*;

(ii)　by *biochemical precipitation*, through organic agencies.

Apart from the direct precipitation of calcite ($CaCO_3$) from seawater during evaporation, calcite (or aragonite, Section 3.2.3) may be precipitated biogenically. Many marine organisms build shells of calcium carbonate by extracting Ca^{2+} cations and HCO_3^- (bicarbonate) anions from seawater. Calcium carbonate is removed, therefore, when the calcareous skeletons of these organisms sink to accumulate as sediments. When the sediments contain predominantly micro-organisms, they acquire the general appearance and grain size of muds and are known as *calcareous oozes*[A]. They occur in many modern marine environments. They may eventually become compacted and cemented into fine-grained sediments such as chalk (a type of limestone).

The direct precipitation of salts from seawater ((i) above) can also result in deposits which, although less abundant than the types discussed above, are important economically. These occur in certain estuaries and enclosed bodies of water that have restricted circulation and interchange with the main mass of ocean water. The salts become concentrated and this favours their direct precipitation. In cases of extreme evaporation, salts other than $CaCO_3$ may be deposited by direct precipitation as *evaporite*[A] deposits. It is at this stage that K^+, Na^+ and Mg^{2+} may form minerals of great chemical purity—some of these minerals are listed in Table 11.

Other economically important deposits that we have not mentioned are *carbon-rich organic materials*, for example coal and oil, that either enter the sedimentary cycle of erosion and deposition from land or accumulate from surface ocean water. (These will be discussed in Block 5.)

Referring again to Table 16, we see that the average composition of continental crust (represented by rocks between granodiorite and diorite in Table 14) has been segregated into quartz-rich, clay-rich and limestone-rich fractions by a combination of physical and chemical selection and reselection processes. The overall process that brings about the accumulation of progressively more pure sedimentary deposits, and that has been in operation since the crust of the Earth first formed, is termed *sedimentary differentiation*.

sedimentary differentiation

It is also important for you to realize that the chemical and physical reactions that produce sediments do not stop immediately after deposition. Physical compaction and chemical deposition of cementing material in the spaces between grains both lead to the production of rocks from sediments: this is diagenesis. In claystones and mudstones, for example, increasing mass of sediments deposited on top causes compaction, encouraging the sheet-silicate clay minerals to become preferentially oriented, leading to *shales* (mudstones that split along bedding planes). At this point, there is a transition between the end of sedimentary processes and the onset of metamorphic processes.

4.3　Metamorphic processes and products

Metamorphism is a term used to describe the *changes that affect pre-existing rocks when they are subject to high pressures and/or temperatures*. It usually refers to chemical and physical reactions that take place in the *solid state*, though many metamorphic reactions are greatly assisted by the presence of a *liquid* or *gas* that permeates through the rock. But metamorphism does not include the production of silicate melts—that is considered to be in the realm of *igneous* processes—although, as you will see, there is overlap between the conditions of igneous and metamorphic processes just as there is between sedimentary and metamorphic processes. Metamorphism is essentially a *rebuilding process* on a mineralogical scale and many metamorphic reactions are endothermic. They also lead to new mineral assemblages of smaller total volume (greater density) than the reactants.

ITQ 28 Given a case in which a reaction is both endothermic and leads to products of lower total volume, will it be favoured by

(a) increasing temperature?

(b) decreasing temperature?

(c) increasing pressure?

(d) decreasing pressure?

Because *pressure* is an important parameter in metamorphism, we define the *change in volume caused by a reaction* (ΔV), *known as volume of reaction*, as:

volume of reaction

$$\Delta V = V_{\text{products}} - V_{\text{reactants}} \qquad (14)$$

You will see that reactions favoured by increasing pressure have negative values of ΔV. The volume of the products is less than that of the reactants (as you might expect with increasing pressure from Le Chatelier's principle). Conversely, reactions favoured by increasing temperature have positive values of ΔH. In Figure 45 you can see the relationship between the ΔH and ΔV of reactions and the slope of the boundary between the pressure–temperature 'fields' where the products and reactants are most stable. If ΔH and ΔV have the same sign, the slope of the boundary is positive; if ΔH and ΔV have opposite signs the slope is negative. Much use has been made of these simple thermodynamic facts in studying metamorphic reactions; you will meet some examples below and, again, later in the Course.

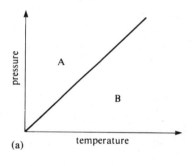

(a)

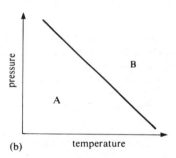

(b)

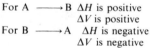

For A $\longrightarrow$ B ΔH is positive
$\qquad\qquad\qquad$ ΔV is positive
For B $\longrightarrow$ A ΔH is negative
$\qquad\qquad\qquad$ ΔV is negative

For A $\longrightarrow$ B ΔH is positive
$\qquad\qquad\qquad$ ΔV is negative
For B $\longrightarrow$ A ΔH is negative
$\qquad\qquad\qquad$ ΔV is positive

FIGURE 45 The relationship between enthalpy and volume changes during reactions involving two mineral phases (or assemblages) A and B. In each case, ΔV and ΔH are considered for the pressure–temperature conditions *at* the boundary curve. You will see that in (a) the same kind of boundary is drawn as that for carbon polymorphs (Figure 24) and that for $CaCO_3$ polymorphs (Figure 54). (Note that H (enthalpy) is not the energy quantity shown in Figure 23 and the two diagrams should not be confused. A more detailed thermodynamic explanation is not possible at this stage in the Course.)

You have already studied some of the features of metamorphism in AV 02. There are three main kinds of metamorphism (summarized in Figure 46):

1 *Contact metamorphism*[A] characterizes the boundary zones of hot igneous intrusions where temperatures become abnormally high because the hot magma heats the rock that it intrudes.

2 *Regional metamorphism*[A] refers to the widespread effects of both increasing pressure and temperature as pre-existing rocks become buried in the crust. The average increase of temperature with depth (or *geothermal gradient*) in the continental crust is about 30 °C km^{-1}; this is indicated by a red line in Figure 46. Regionally metamorphosed rocks may be revealed at the Earth's surface by the fracturing and folding of the crust (for example, during mountain building resulting from continental plate collisions) and by the erosion of higher layers of less metamorphosed materials.

geothermal gradient

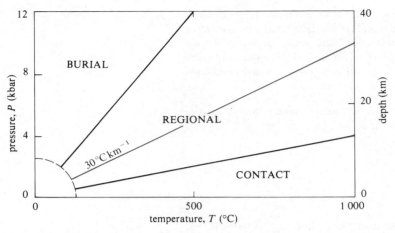

FIGURE 46 Schematic illustration of the pressure–temperature conditions for different kinds of metamorphism. (The red line follows a thermal gradient of $30°C\,km^{-1}$.)

3 *Burial metamorphism* occurs at high pressures and *low* temperatures. The commonest environment for such conditions is in, or bordering, subduction zones where crustal materials are faulted down to high pressures before they have a chance to heat up. If they find their way back to the surface, by further collision events, for example, the products we observe are characteristically low-temperature, high-pressure assemblages.

burial metamorphism

> Look again at Figure 46. What kind of geothermal gradients characterize contact and burial metamorphism? (To work out these gradients, you need to divide the value of temperature at a particular depth in each zone by the depth in kilometres.) How do they compare with those of regional metamorphism?

Geothermal gradients associated with contact metamorphism ($c.\ 100°C\,km^{-1}$) and burial metamorphism ($c.\ 10°C\,km^{-1}$) are, respectively, higher and lower than in regional metamorphism.

You may be wondering how we *know* that the pressure–temperature conditions given in Figure 46 characterize these different kinds of metamorphism. The answer is that the rocks that have resulted from these processes have various distinguishing features, such as their mineralogy, that can be studied through laboratory experiments at high pressures and temperatures and by using thermodynamic calculations. In these ways the conditions of production for metamorphic rocks may be deduced, giving the picture in Figure 46.

Let us now consider what happens to different kinds of sediments when they are metamorphosed. Sandstones and limestones are nearly pure quartz and calcium carbonate respectively, and we can start with them. When they are heated, they mostly just recrystallize, to produce *quartzites* and *marbles*. But if there are other constituents in the rocks, then chemical reactions take place. We shall begin with a simple example.

quartzite   marble

> **ITQ 29** Take the case of an impure limestone that contains quartz grains (Figure 47a). In the marble that is produced by metamorphism (Figure 47b), small needles of the mineral wollastonite ($CaSiO_3$) are sometimes found within the grains of quartz. The reaction that has taken place is:
>
> $$CaCO_3\ +\ SiO_2\ =\ CaSiO_3\ +\ CO_2$$
> limestone   quartz   wollastonite   gas
>
> (a)   Is this reaction exothermic or endothermic?
>
> (b)   Does it lead to an increase or decrease in volume?
>
> (c)   Does it have a positive or negative slope in a pressure–temperature diagram?

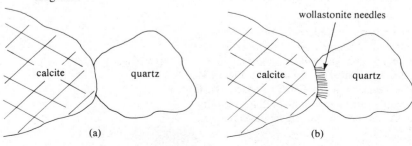

FIGURE 47 Sketch to show the formation of wollastonite ($CaSiO_3$) needles within a quartz grain during metamorphism of an impure limestone. The two diagrams (a) and (b) show the situation before and after metamorphism.

So impure limestone and sandstones can grow new minerals, such as wollastonite. Earlier, we discussed the relatively impure and mixed nature of mudstones (see Table 16). They are, in fact, prone to very great changes during metamorphism on account of their chemical complexity. First, the textural changes: you know that clays are minute flakes of sheet-silicate structures. These are deposited in random orientations in most muds. But, on compaction, the minute flakes get rotated round to be parallel with the layering and at right angles to the compacting load. When this compaction is accompanied by heating, the clay mineral particles begin to recrystallize and the process is promoted by hot water-rich fluids (often also containing dissolved gases such as CO_2) being forced through the sediments at high pressures caused by the compacting load of overlying sediments. In this way, new micaceous (mica-rich) minerals form. These minerals are sheet-like or platey and they respond by becoming oriented perpendicular to the principal stress acting on the rocks and causing the metamorphism. This stress is usually in a vertical direction due to the compacting load, but often varies from this in zones of crustal shortening due to plate movements. Consequently, in such highly deformed and folded zones of regional metamorphism where the rocks are contorted, this principal stress is not always parallel to the original layering.

In *slates*[A], because of their origin as clays, the minerals are still microscopic, but they are also largely micaceous: their preferred orientation within the rock and ability to cleave often gives the entire rock a well-defined *cleavage* (exploited in the manufacture of roofing slates, for example). More extensive regional metamorphism of mudstones results in larger flakes of mica: coarsely crystalline rocks that contain these are called *schists*[A] (see specimen 9 in your Home Experiment Kit), a term that refers to the texture of foliated* rocks and that is not confined exclusively to metamorphosed mudstones. But micas are not the only minerals that form during metamorphism. Garnets form as the pressure rises, and other minerals with dense and compact structures are produced by reactions with negative ΔV. We shall look at some examples in a moment.

In *contact metamorphism* of mudstones, where temperature is the dominant control, new minerals such as andalusite (Al_2SiO_5, discussed below) may grow as spots and, if the rock has already developed a fabric of aligned clay minerals (as described above), it takes on the appearance of a *spotted slate* (*CB*, Plate 25a). Closer to the intrusion causing the contact metamorphism, where temperatures are higher, more extensive new mineral growth may produce hard, banded rocks called *hornfelses* (*CB*, Plate 25b). The banding recognized in hand specimens is due to the segregation of light and dark minerals. Because the type of contact metamorphic rock varies with the intensity of metamorphism, concentric zones are often found around intrusions. These zones may be characterized by hornfelses, spotted slates and then, with increasing distance from the contact, unaltered shales or mudstones.

hornfels

From what we have said so far, it should be clear that, in most cases of metamorphism, there is a change in mineralogical but not in chemical composition. Metamorphic changes are promoted, or catalysed, by the associated liquid (and/or gas) which permeates through the rock. Because there is virtually no change in chemical composition (the process is almost isochemical), the typical analyses given for the sedimentary rocks (sandstone, mudstone and limestone) in Table 16 could apply equally well to their metamorphic products: quarzites, slates and schists, marbles. Interestingly, metamorphism works in the *opposite* sense to the chemical breakdown that we described for chemical weathering. On an atomic scale *minerals are rebuilt*: clay minerals may be the starting point for the metamorphism of mudstones but they are rebuilt into feldspars, micas, amphiboles and other metamorphic minerals.

The range of metamorphic products starting with mudstones may seem pretty extensive. But a different range of metamorphic rocks results from the metamorphism of igneous rocks at high pressure and temperature, and this is partly because igneous rocks have a wide range of chemical composition. The details of all these changes need not concern us here, but you should note that the presence of a water-rich fluid is important. For example, metamorphosed basalts take on a green appearance, due to the incorporation of iron (Fe^{2+}) into a sheet-silicate mineral, chlorite: they produce *greenschists*. At higher pressures and temperatures, metamorphosed basalts may yield feldspar and amphibole-rich rocks, like the *amphibolite gneiss* (specimen 10) in your Home Experiment Kit. (*Gneiss*[A] is a textural term that refers to coarsely banded metamorphic rocks which may have various compositions—see Table 12 and *CB*, Plate 27, for example.)

* Foliation describes the aligned micaceous texture of certain rocks.

The accumulated evidence of metamorphic changes in all kinds of rock types has led to the concept of *metamorphic facies* (labelled in Figure 48): *a facies is defined as a set of mineral assemblages that occurs repeatedly in space and time, that can be related to a particular set of pressure–temperature conditions during metamorphism and that depends also on gross chemical composition.* Each set of mineral assemblages is, therefore, characteristic of a particular range of pressure–temperature conditions.

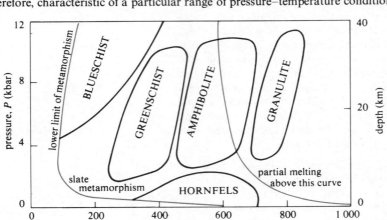

FIGURE 48 The pressure–temperature conditions for the major metamorphic facies. These are defined by convention using the products of basalt metamorphism and are used for convenience to describe the pressure–temperature conditions of metamorphism for all other rock types. (The boundaries shown encompass the P–T conditions under which each facies assemblage is most commonly produced; in fact, the boundaries are often blurred due to solid-solution effects.)

Because of the impressive way in which metamorphosed basalts develop different metamorphic textures and mineralogy, it is customary to define the facies in terms of these products (see Figure 48). Apart from the familiar silicates such as feldspars, micas and amphiboles, some of the new minerals that grow during regional metamorphism are particularly diagnostic of the conditions in which they formed. A particularly useful example is provided by three minerals that occur as minor components in many rocks of the greenschist and amphibolite facies rocks: the *aluminosilicates*.

All three minerals have the same chemical formula, Al_2SiO_5, but their internal structures and external appearances are markedly different (*CB*, Plate 26). Can you remember what the minerals belonging to such a group are called?

Minerals that have identical chemistry but different structures are polymorphs (Section 3.2.3). Each aluminosilicate is typical of a different range of pressure–temperature conditions and the boundaries where one mineral turns into another (polymorphic transition boundaries) are shown in Figure 49. The three polymorphs are called kyanite, andalusite and sillimanite.

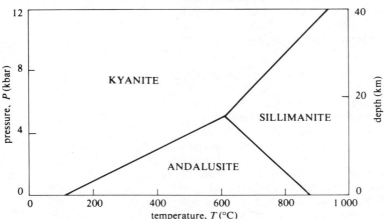

FIGURE 49 The boundaries between the stability fields of the three aluminosilicate minerals in pressure–temperature space.

ITQ 30 (a) Look at Figure 49. Can you place the aluminosilicate minerals in order of increasing enthalpy and increasing volume?

(b) Which of these three polymorphs will occur most commonly in contact metamorphism, and which will occur most commonly in the amphibolite facies of regional metamorphism as defined in Figure 48?

As you can see from Figure 49 and ITQ 30, aluminosilicates are potentially very useful in crudely dividing the physical conditions under which metamorphic rocks formed into high-pressure (kyanite), low-pressure (andalusite) and high-temperature (sillimanite) regimes. Other minor minerals can be used in similar ways and sometimes allow the conditions under which they formed to be pin-pointed quite accurately.

We have commented in some detail on the common products of contact and regional metamorphism, but have said little about *burial metamorphism*. Some of the most spectacular products occur in metamorphosed basalts, which we think have been carried down into a subduction zone but then brought back to the surface by large-scale faulting or folding. Metamorphosed basalts in the blueschist facies of metamorphism (Figure 48) are dominated by garnets and spectacular blue–green amphiboles—both are low-volume, *dense* minerals and occur in eclogite, which is specimen 8 of your Home Experiment Kit.

Finally, in this brief introduction to metamorphism, we will consider the source of the fluids involved. Take the example of a mudstone which progresses through the pressure–temperature conditions of the greenschist and amphibolite facies and which, eventually, reaches the granulite facies.

> Can you see where the fluid that accompanies metamorphic changes comes from in such a case?

Metamorphism can also be thought of as a dehydration process. As clay minerals are converted back to more compact silicates they release their water—though some is retained as hydroxyl groups in micas and amphiboles. This fluid is free to migrate through the rock pile and along the rock fabric down the pressure gradient (that is, towards areas of low pressure—normally, towards the surface) and, on the way, it may assist with cation exchange between rock types, possibly leading to slight changes in their bulk chemical compositions. The net result of the dehydration process is that the highest grade metamorphic rocks, in the *granulite facies*, contain the least structurally bonded water: such rocks are dominated by anhydrous minerals such as feldspars and pyroxenes. Most of the rocks that have reached these conditions have the coarsely banded appearance of *gneisses* when they reach the surface: they are often called *granulites* or granulite gneisses.

granulite

From Figure 48 you can also see that the transition between the upper amphibolite and lower granulite facies overlaps with the beginning of partial melting for crustal rocks. Thus, the most extreme conditions of metamorphism may be associated with the generation of melts in the crust. Such melts are known from experiments to be acidic and to yield granite bodies after they have risen and crystallized. The partial melting of crustal rocks and the ascent of aqueous fluids combine to cause chemical change and to distil alkali elements and silica upwards, even *within* the crust. We shall return to this theme in later Blocks (particularly Blocks 2, 4 and 7). But, in the meantime, we would like you to remember that *metamorphism is an essentially isochemical process that leads to the growth of new minerals, often with upwards expulsion of fluids, and, at the highest grades of metamorphism, partial melts in the form of acid magma.*

4.4 Summary of Section 4

This introduction to the processes of igneous activity, erosion and sedimentation, burial and metamorphism, melting and further igneous activity, should have proved to you that the chemical elements are continuously recycled *within* the continental crust. Through magmatism, the bulk of the crust continuously increases above subduction zones, but some magmas are derived from within the crust and are an integral part of chemical recycling. All these processes are various stages in the *rock cycle*[A], which is summarized in Figure 50. From below, the cycle is driven by the Earth's internal heat (you will learn more about this in Block 4), and this heat is dominantly responsible for melting and metamorphic processes. The surface part of the cycle is driven by *solar energy* which is responsible for the climatic controls on weathering processes. The chemical elements are continuously sorted and resorted by all the components of the cycle. In physical terms, the cycle has been summarized as follows:

> In many places igneous activity is related to deep-seated fractures; these fractures may act as channels for gases carrying heat from greater depths, and movements on such fractures will disturb the balance between temperature and pressure on either side. Such mechanisms may produce magma, which by virtue of its lower density than the surrounding solid material tends to rise into the higher and cooler parts of the crust. The thermal energy of the magma is dissipated to the surroundings during cooling and solidification. The kinetic energy of the atoms or ions decreases as the temperature falls; their potential energy also diminishes when they are arranged in orderly fashion in crystals. At the next stage of the cycle erosion and sedimentation reduce the material on which they work to a lower energy state. Erosion moves material with the aid of gravitational forces, and a loss of potential energy results. The chemical reactions accompanying weathering are spontaneous and usually irreversible and lead to a decrease in energy. The general trend is reversed during metamorphism, when thermal and gravitational energy is converted into chemical energy by endothermic reactions and the formation of compounds of higher density. A sufficient increase in temperature may increase the kinetic energy of the atoms or ions sufficiently to overcome the forces holding them in crystal lattices, so that the minerals decompose or melt. A magma is thereby regenerated, and the cycle is complete.
> (B. Mason (1966), *Principles of Geochemistry*, John Wiley & Sons, Inc.)

In the wider context of the Earth's structure, composition and evolution, we showed earlier in this Block that the lithophile part of the Earth comprises the crust and mantle: most of this is peridotitic (iron–magnesium silicate). The surface rocks of the continental crust, most extensively studied by geologists, are a multi-staged

FIGURE 50   The rock cycle.

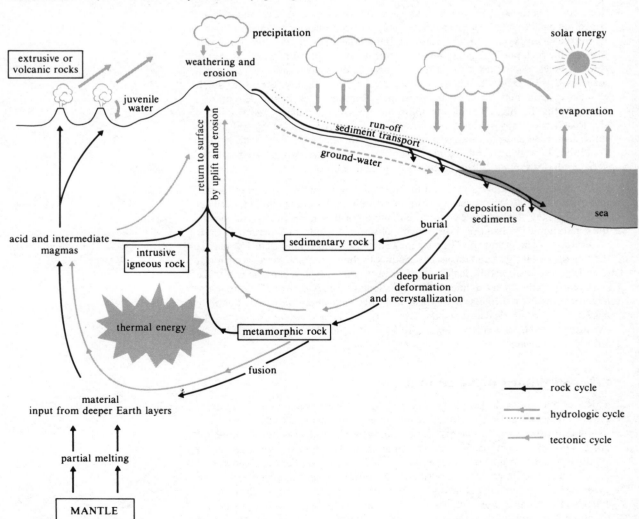

distillate of mantle peridotite which incorporates a few per cent of its most easily melted components. On average, these rocks have chemical compositions between diorite and granodiorite: they are enriched in silica, alumina, alkalis and other very lithophile elements compared with the peridotite mantle. Physical and chemical processes at the surface further differentiate this chemical mixture and, in some cases, dramatically increase the purity of the products (for example, in sandstones and evaporites). In terms of chemical purity these rocks may be regarded as end products of the rock cycle, but they are not immune from further reprocessing. There is, however, some evidence from the geological record that sedimentary differentiation has become more effective during the Earth's history. This suggests that, like the progressive growth of continents, there may be an irreversible trend in the chemical reprocessing of the Earth's continental crust.

In Block 2 we shall introduce you to some of the physical evidence from earthquakes, gravity measurements, etc., that constrains models for the internal composition of the Earth and, at the end of that Block, we shall refine our knowledge of the composition of the Earth's layers. Meanwhile, we suggest that the most fruitful way of revising Section 4 of this Block is to use Figures 41, 42, 44, 48 and 50 in conjuction with this summary and the tape for AV 02.

4.5   Objectives for Section 4

Now that you have completed Section 4 you should be able to:

1   Recognize valid definitions of the terms flagged in the margins of this Section.

18   Critically evaluate geochemical data obtained by sampling, averaging and from chemical analysis, and carry out simple statistical calculations using different geochemical data (Appendix 1).

19   Explain the principles used in geochemical analysis by (a) wet chemistry; (b) spectrochemistry; (c) mass spectrometry; (d) X-ray fluorescence; and (e) neutron activation analysis; and evaluate the relative precision, accuracy and sensitivity of each method (Appendix 1).

20   Identify, with reasons, the mineral content of various igneous, metamorphic and sedimentary rocks, and link this to the major element chemical compositions of such rocks.

21   Use the parameters of enthalpy and volume for minerals involved in reactions to deduce the broad physical conditions under which they are most stable.

22   Account for the different temperatures of crystallization and different resistances to chemical breakdown of the primary silicate minerals, and give examples of their behaviour and fate in magmatic, metamorphic and surface processes.

23   Indicate a broad understanding of where different major rock types are found in the Earth's crust and relate this to the processes of their origin.

24   Show how, by the influence of the dynamic internal and surface processes of the rock cycle, the Earth's continental crust has become chemically differentiated from, and enriched in fusible components relative to, the mantle.

Apart from Objective 1, to which they all relate, the eleven ITQs in this Section and Appendix 1 test the following Objectives:

ITQ 23, Objective 20; ITQ 24, Objectives 15 and 20; ITQ 25, Objectives 15 and 20; ITQ 26, Objective 22; ITQ 27, Objectives 16 and 22; ITQ 28, Objective 21; ITQ 29, Objective 21; ITQ 30, Objectives 20 and 21; ITQ 31, Objective 18; ITQ 32, Objective 18; ITQ 33, Objectives 10 and 19.

Now try the following SAQs, which test other aspects of these Objectives.

SAQ 13 (*Objective 18*)   Which of the methods (i)–(iv) would you use to process the samples for the problems in (a)–(d)? (Select two alternatives in each case. You may need to refer to Appendix 1.)

(i)   Random sampling

(ii)   Systematic sampling

(iii)   Analyses that are accurate and precise

(iv)   Precise analyses only

(a)   Samples of soil from Cornwall taken to detect the areas richest in tin.

(b)   Samples of a large number of minerals analysed for K and Rb in order to determine whether a relationship exists between the amounts of these two elements.

(c)   Samples of metamorphic rocks representative of a particular area in northern Scotland to determine the average potassium content of the crust in this region.

(d)   Samples of mica from different rock types taken to calculate their average compositions in the different types.

SAQ 14 (*Objective 20*)   The chemical and/or mineralogical information given in (a)–(d) applies to four common rocks, and you are to identify, in each case, the most likely rock type involved. Where the mineralogical composition is given, comment on the probable major element chemistry of the rock, and vice versa. (Use Tables 12, 14 and 16.)

(a)   A coarse-grained homogeneous rock containing 10 per cent quartz, 70 per cent feldspars and 20 per cent amphibole.

(b)   A coarse-grained monomineralic rock (that is, containing just one dominant mineral) containing 40 per cent SiO_2 and 45 per cent MgO.

(c)   A fine-grained monomineralic rock containing 35 per cent CaO, 15 per cent MgO and 40 per cent CO_2.

(d)   A coarse-grained rock containing cemented fragments of quartz (80 per cent) and alkali feldspar (20 per cent).

SAQ 15 (*Objectives 20 and 21*)   (a)   Look at Figure 54 (p. 93). Has calcite or aragonite

(i)   the higher enthalpy?

(ii)   the higher volume (lower density)?

(b)   From the following list (A–D) what are the most likely regional metamorphic rock equivalents of

(i)   basalt?

(ii)   sandstone?

A   hornfels   B   quartzite   C   amphibolite   D   marble

SAQ 16 (*Objectives 23 and 24*)   Are the following statements true or false? Give reasons for your answers.

(a)   The process in which basalts are metamorphosed to blueschists is known as contact metamorphism.

(b)   The process of continental crust formation has depended increasingly on solar energy during the Earth's history.

(c)   The development of a fluid phase during regional metamorphism is often due to progressive dehydration of clay minerals with increasing temperature.

(d)   The average composition of the ocean crust is that of basic rocks, whereas that of the continental crust is between those of intermediate and acid rocks.

SAQ 17 (*Objectives 22, 23 and 24*)   (a)   Look at Table 18 (answer to ITQ 23). Can you account for:

(i)   the lack of ferromagnesian minerals in many sedimentary rocks, other than Fe^{2+} and Mg^{2+} in clays (Table 16)?

(ii)   the presence of feldspar and/or clay minerals in different sedimentary rocks?

(iii)   the presence of ferromagnesian minerals in metamorphosed clay-rich sediments (greenschist and amphibolite facies)?

(iv)   the presence of garnet in metamorphic rocks formed at high pressures?

(b)   Trace the history of sodium through the rock cycle from its primary source in mantle peridotite to a repository in salt deposits.

Optional further reading

General

Mason, B. (1966) *Principles of Geochemistry*. John Wiley and Sons, Inc., London.

Solar System and stellar evolution

Mitton, S. (ed.) (1977) *The Cambridge Encyclopaedia of Astronomy*. Trewin Copplestone Publishing Ltd., London.

Scientific American, September 1975 (articles on most planets included).

Wood, J. A. (1969) *The Solar System*. Prentice-Hall, Inc., Englewood Cliffs, New Jersey.

Geochemical classification and relevance to Earth of stellar evolution

Brown, G. C. and Mussett, A. E. (1981) *The Inaccessible Earth*. George Allen and Unwin, London.

Turekian, K. K. (1971) *Chemistry of the Earth*. Holt, Rinehart and Winston, Inc., New York.

Minerals and rocks

Cox, K. (1972) 'Minerals and Rocks' In Gass, I. G., Smith, P. J. and Wilson, R. C. L. (eds.) *Understanding the Earth*, 2nd edn., Artemis Press, Sussex.

Ernst, W. G. (1969) *Earth Materials*. Prentice-Hall, Inc., Englewood Cliffs, New Jersey.

Appendix 1 Geochemical analysis

The chemical analyses given in Tables 13b, 14 and 16 total around 100 per cent, within the experimental error of the analytical techniques (see below), when only 10–12 so-called *major elements* are considered. But there are many other naturally occurring elements which, apart from their concentration in ore deposits, for example, are commonly dispersed in *minor* (small, say, <0.5 per cent) or *trace* (very small, say, <0.05 per cent) amounts throughout most rocks. Many elements can be detected in rocks in abundances of a few parts per million (p.p.m.; 100 p.p.m. = 0.01 per cent) and the distinction between trace, minor and major elements is often *quite arbitrary*. Some trace element concentrations are found to fluctuate in geological systems to a much greater degree than major elements and this makes them very useful for studying problems that are difficult to solve by conventional methods. Block 3 gives some examples.

major elements

minor elements   **trace elements**

Reliability of geochemical data

The analyses in Tables 13b, 14 and 16 are *averages* obtained by summarizing large amounts of data in a single analysis. At present, you have no means of knowing how the averages were calculated, how reliable they are, or how much variation they conceal. There are several important factors that must be considered if we are to determine the reliability of geochemical analyses: here we give them a brief airing; note that they do not form a major part of this Course.

1 *Sampling pattern* Suppose you wished to obtain a set of analyses and a representative average for a single granite intrusion, several km in diameter, there are two possible approaches:

(a) *Random sampling* You collect samples from random geographical locations within the intrusion. Such a collection of samples does not provide uniform coverage of the area, and so is subject to bias. However, practical considerations, such as the area of available rock outcrop, may dictate the use of random sampling (because outcrops are rarely distributed regularly over an intrusion).

random sampling

(b) *Systematic sampling* You collect samples in a well-defined pattern—for example, at the intersection points on a square grid. Because such a grid can be used to cover a particular area, systematic sampling is the most effective method of collecting samples for the detection, evaluation and interpretation of geochemical variation.

systematic sampling

2   *Sample size* is also important. *The coarser the grain size of the rock, the larger the sample you need.* In fact, there are statistical calculations that can be used to determine the sample size required, in relation to grain size, for a specified reliability of the final analysis. In practice it is usual to collect 2–3 kg of rock sample from the field and to crush this to a fine powder in order to homogenize the sample (that is, to remove chemical variations due to non-random distributions of large crystals). Small portions of powder are then taken for chemical analysis. This technique generally ensures that errors due to sample size are small in comparison with the size of possible laboratory errors (see 3).

3   *Laboratory errors*   In the laboratory, there are three possible sources of error: variable *precision, accuracy and sensitivity*. In this context, *precision* refers to the '*reproducibility*' of an analytical result and it reflects both the skill of the analyst and the reliability of the technique. For example, Figure 51 shows the results of a large number of FeO determinations by the same analyst from the same powdered granite sample. As you may recall from the Science Foundation Course, this type of curve is known as a normal *frequency distribution curve*[A] and the value of FeO beneath the peak of such a symmetrical curve is the *mean value*[A] ($\bar{x}$). The mean value, in this case 2 per cent FeO, could have been calculated from all the observations by dividing their sum by the number of observations (n):

$$\bar{x} = \frac{x_1 + x_2 + x_3 + \cdots + x_n}{n}$$

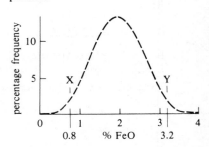

FIGURE 51   The distribution of FeO determinations resulting from multiple analyses of the same granite sample. The points X and Y indicate the 95 per cent confidence limits about a mean value of 2 per cent FeO for this rock.

The *spread* of the values in Figure 51 about the mean is a measure of the precision of the analyses. To obtain a numerical estimate, first we calculate the *standard deviation*[A] (s) by working out the difference (d) of each value (x) from the mean ($\bar{x}$) and by finding the root mean square of these differences:

$$s = \sqrt{\frac{d_1^2 + d_2^2 + \cdots + d_n^2}{n}}$$

Statistical theory shows that 68 per cent of the readings on a normal frequency distribution curve fall within $\pm s$ of the mean, 95 per cent within $\pm 2s$ and 99.7 per cent within $\pm 3s$. These percentages are sometimes used to determine *confidence limits* for a set of analyses; thus, the points X and Y on Figure 51 are at $\bar{x} \pm 2s$ and define the 95 per cent confidence limits. What this means is that if we analyse one sample repeatedly, the chances of any one analysis falling between 0.8 and 3.2 per cent FeO are 95 per cent (95 times out of 100).

More usually, the numerical expression of precision is defined by the *coefficient of variation* (C_v), which is a simple ratio of standard deviation to mean value, expressed as a percentage:

$$C_v = \frac{100s}{\bar{x}} \text{ per cent}$$

As a guide, modern analyses for major element oxides may give values for C_v within 5 per cent, and physical methods for the analysis of trace elements often reach values near 1 per cent. As an example, consider a case where $C_v = 5$ per cent and $\bar{x} = 100$ p.p.m.; the standard deviation would be 5 p.p.m., so the 95 per cent confidence limit of error due to precision would be 100 ± 10 p.p.m., that is, $\bar{x} \pm 2s$. Now check your understanding of these statistical quantities by doing the following ITQ.

ITQ 31   Four analyses of a granite for lead gave results of 19, 20, 17 and 21 p.p.m.

(a)   What are the mean ($\bar{x}$), standard deviation (s) and coefficient of variation (C_v)?

(b)   What is the expected range of lead values in 95 per cent of granite samples analysed by this method?

(c)   Comment on the precision of these analyses.

Accuracy defines the extent to which analytical data *approach to the true value*. In other words, the calculated mean value $\bar{x}$ (Figure 51) for one laboratory may not be the same as $\bar{x}$ in another laboratory and both may differ from the elusive 'true value'. Although accuracy is difficult to assess numerically, attempts have been made to improve inter-laboratory comparisons over recent decades. In the main, these have consisted of the preparation and circulation to all geochemical laboratories of *international standard rock powders*. Organizations such as the United States Geological Survey have been involved in both the circulation of such samples and the collation of results from laboratories world-wide, so enabling each laboratory to compare its results with those of everyone else. *In a nutshell, accuracy is most subject to systematic (constant) errors, whereas precision is affected by random errors.*

Sensitivity refers to the fact that all methods of analysis have a 'lower limit of detection' below which a weak signal (for example, the deflection on a balance needle, the colour of an unknown sample, etc.) cannot be distinguished from the background. Obviously, the effect of sensitivity may drastically reduce the precision or accuracy of an analysis. You should, by now, have little difficulty in understanding that *precision, accuracy and sensitivity are all interrelated*: none can be considered in isolation.

sensitivity

This survey of factors affecting the reliability of geochemical analysis highlights the difficulty of producing geochemical data that are comparable between different field locations and laboratories. It is usually possible to allow for field-based sampling size and distribution errors and even to make some allowance for laboratory precision and inter-laboratory accuracy using international standards. When all these influences have been taken into account, there remain the geological factors that controlled the chemical evolution of the rock before it was sampled. Of course, these are the factors that the geochemist seeks to identify and interpret by analysing suites of samples and such factors are discussed in the Main Text of this and later Blocks. The aim of this discussion was to give you some idea of the practical factors that affect rock analyses.

Analytical techniques

In Section 2.2.1 we introduced the technique of *spectrochemical analysis*, which is just one of the many *physical methods* of chemical analysis in common use. The techniques we shall introduce in this Section use either physical or chemical principles; they are intended to give you an idea of the *range* of possible methods— you will meet other techniques later in the Course. First we shall mention some techniques that are simply *chemical* in their operation.

Chemical methods

These are often termed '*wet chemical methods*' because the rock has first to be dissolved. For limestones, this is quite simple because they dissolve in dilute hydrochloric acid, but most rocks containing silicates must be dissolved in strong hydrofluoric acid. The solution produced can be analysed in several different ways. The most *accurate* method of analysis, which does not require a standard for comparison, is *gravimetric*, but this is often a slow tedious operation. It involves precipitating insoluble compounds from solution, weighing them, and calculating element concentrations from the mass and known compositions of these precipitates.

gravimetric analysis

> **ITQ 32** 200 mg of an igneous rock sample is dissolved and the calcium in the solution is precipitated as calcium oxalate, $Ca(COO)_2$. This is heated strongly to give 20.8 mg of calcium oxide, CaO. What percentage of CaO can be ascribed to the analysis of this rock and what rock type might this be?

There are also several more rapid *comparative* methods of wet chemical analysis that avoid precipitation and weighing operations. Many of these require reagents to be added to the rock solution under closely controlled conditions, to form *coloured compounds* of specific elements.

The intensity of the resultant colour is proportional to the concentration of the element and intensities are measured on a *spectrophotometer. Colorimetric* methods are much more rapid than gravimetric methods, but they require some chemical manipulation and may seem tedious in comparison with purely physical methods. However, they often have very high absolute accuracy.

colorimetric analysis

Physical methods

(a) *Mass spectrometry*[A] All you need to know about mass spectrometers here is that they are instruments that are capable of *ionizing atoms* of all elements which are then *sorted according to their nuclear masses and ionic charges*. The method is particularly suited to the radiometric dating of rocks (Block 3) because it determines *isotope ratios.* To obtain quantitative data, it is necessary to add carefully controlled amounts of standard material to the sample being analysed. The sensitivity of mass spectrometry can be very high (10^{-6}–10^{-12} g, meaning that less than 1 p.p.m. of a particular isotope can be identified in a gram of sample)* and is much higher than in emission spectrometry, where the lower limit of detection may be as high as 100 p.p.m. On the other hand, mass spectrometry is a relatively slow technique, but it usually has a high absolute accuracy (1–5 per cent) and precision ($C_v = 1$–3 per cent at levels down to 1 p.p.m.).

(b) *X-ray fluorescence spectrometry* Instead of using heat to excite the atoms of the material (as in emission spectrometry) *an X-ray beam* is used (red in Figure 52a). In this case, the greater energy input causes electrons *near the nucleus* to be excited and then to emit secondary, or *fluorescent X-rays* on reverting to their original energy states. The short wavelength X-rays ($\lambda = 10$–$1\,000$ pm) are sorted by diffraction in a pure *analysing crystal* of known *d*-spacing (from the Bragg equation $n\lambda = 2d\sin\theta$, see Section 3.2, Figure 17).

(a)

(b)

FIGURE 52   (a) Photograph of the working parts of an X-ray fluorescence spectrometer. (b) Diagram of the essential working parts—see Figure 17 for definition of θ. (Note that a collimator is a device, usually a metallic grid, designed to focus the X-ray beam.)

Because we know *d* for the crystal very accurately, we can set θ to any desired value. Radiation will then be collected at the detector for a unique wavelength which is characteristic of the element being analysed. The intensity of radiation is measured, relative to a standard treated in the same way, and is proportional to the concentration of the element being excited.

> **ITQ 33**   For the analysis of silicon, an X-ray emission line ($\lambda = 713$ pm) is used with an analysing crystal of *d*-spacing $= 437$ pm. By using the Bragg equation, calculate the angle, θ, at which the analysing crystal must be set to diffract the first-order diffraction line ($n = 1$, Section 3.2) into the detector.

The usual method of obtaining X-ray fluorescence (XRF) results is, first, to draw a calibration graph of X-ray intensity against concentration for standards, and, second, to refer the X-ray intensities of the unknowns to this graph. XRF methods are much more rapid and flexible than mass spectrometry and are often used to analyse large numbers of samples for a wide range of elements (but not isotopes).

*This is equivalent to selecting a single individual in the population of the United Kingdom on the basis of weight!

The sensitivity of the technique, for most elements, is less than 5 p.p.m. Accuracy is usually quite good, depending on the quality of the standards employed, and C, is usually within 5 per cent.

(c) *Neutron activation analysis* This is one of the most recently developed physical methods and it depends on the 'activation' of the elements of interest by bombardment of the sample with *fast-moving neutrons* in the core of a nuclear reactor. *Activation*, in this context, means the addition of neutrons to the nucleus of stable isotopes to produce new radionuclides. These newly produced isotopes then decay, producing *particles with characteristic energies* that can be recognized and measured with a scintillation counter. Many samples are packed into a single reactor capsule in a specified geometrical fashion together with standards. In this way, the variation in activation efficiency across the capsule is determined as are the element concentrations in the unknowns.

For example, the concentration of uranium may be determined by activating the common isotope ^{238}U to ^{239}U and then measuring its decay:

$$^{238}_{92}U \xrightarrow{\text{+ 1 neutron}} {}^{239}_{92}U \xrightarrow{\beta^- \text{ decay}} {}^{239}_{93}Np \xrightarrow{\beta^- \text{ decay}} {}^{239}_{94}Pu$$

The two β^--decays of ^{239}U and ^{239}Np are, of course, time-dependent. They have half-lives of 23.5 minutes and 2.35 days respectively and so measurements must be made soon after activation. You may recall that ^{238}U, the parent isotope in this case, is also radioactive but, with a half-life of 4 467 Ma, it can be regarded as stable for the purpose of activation analysis.

Although time-consuming, and requiring reactor facilities, activation analysis has better sensitivity (often <1 p.p.m. detection limits) than most other physical methods of rock analyses. Precision is comparable with other methods but there are a few unanswered questions about accuracy, largely due to the uncertainties of standardization.

To summarize: despite the continued improvement and diversification of physical analytical techniques, it is unlikely that wet chemical methods will ever be entirely superseded. Chemical methods are lengthy but, for many elements, are the only way in which really accurate data can be obtained in order to standardize physical methods of analysis.

Appendix 2

Home Experiment Kit rock samples

You should refer to this after completing AV 02.

The following brief petrological and mineralogical descriptions should be compared with the information given in Figures 41, 46 and 48.

1 *Granite* (Peterhead, Scotland) Coarse-grained, equigranular crystalline rock comprising pink grains of feldspar (showing rectangular cleavage faces), grey, glassy quartz (irregular outlines) and occasional flakes of black biotite mica.

2 *Microdiorite* (northern Italy) Medium-grained, equigranular crystalline rock dominated by feldspar (white matrix) together with black needles of amphibole and shiny black flakes of biotite mica.

3 *Gabbro* (north-west Scotland) Coarse-grained, equigranular crystalline rock with roughly equal amounts of feldspar (white or greenish-white, frequently with rectangular crystal outlines) and pyroxene (black, also rectangular).

4 *Peridotite* (Norway) Medium-grained, equigranular crystalline rock almost entirely composed of green to grey–green olivine crystals; note this sample's greater density than that of samples 1–3.

5 *Porphyritic flow-banded rhyolite* (Glen Etive, Scotland) Fine-grained crystalline rock with alternating light and dark brown bands resulting from the streaking out of highly viscous blobs of lava, with slightly variable compositions, before the rock cooled. Occasional larger rectangular feldspar phenocrysts (several mm in length) must have formed before the groundmass crystallized.

6 *Porphyritic andesite* (central France) Medium brown, fine-grained crystalline groundmass with small holes (vesicles) due to escape of trapped gas bubbles from the

cooling lava. Two phenocryst minerals can be distinguished: fine white needles of feldspar ($c.\ 1 \times 5$ mm) are aligned on some rock surfaces, indicating the direction of flow; larger ($c.\ 3 \times 5$ mm) irregular black phenocrysts of amphibole or pyroxene (the latter in thin section) are also abundant.

7   *Porphyritic basalt* (central France)   Fine-grained, dark grey crystalline groundmass with larger ($c.\ 1$ cm), abundant phenocrysts of black pyroxene (with occasional rectangular outlines) and smaller ($c.\ 3$–4 mm), less abundant phenocrysts of brown–green olivine.

8   *Eclogite* (Genoa, northern Italy)   A crystalline rock of great density comprising roughly equal amounts of interlocking dark green crystals of pyroxene and discrete, round, red–brown crystals of garnet. Some specimens carry layers of silvery muscovite mica. The rock is the result of burial metamorphism of basic igneous rocks and, for our purposes, may be equated with the *blueschist* facies of metamorphism in Figure 48.

9   *Mica schist* (Inverness, Scotland)   Medium- to coarse-grained rock dominated by white (muscovite) and dark (biotite) mica in sheet-like crystals which are all aligned to form the foliated appearance typical of schists. Other minerals are white feldspar and occasional elongate, pale blue crystals of kyanite (see Figure 49). This rock is the result of regional metamorphism of mudstones and formed in the greenschist facies of metamorphism in Figure 48.

10   *Amphibolite gneiss* (north-west Highlands of Scotland)   Coarsely banded rock comprising light grey bands of finely crystalline feldspar and black bands of finely crystalline amphibole. This rock is the result of regional metamorphism of basic igneous rocks and formed near the amphibolite–granulite facies boundary in Figure 48.

11   *Fine-grained sandstone* (Penrith, Cumbria)   Most of the grains in this rock are quartz, but their glassy appearance is obscured by cementing overgrowths of red–brown iron oxides (released by intense chemical weathering of pre-existing rocks in desert conditions). The grains are well rounded and well sorted, but some variation is visible as bedding. The less abundant, small light brown grains in this specimen are mainly feldspar.

12   *Greywacke* (Yorkshire Dales)   This rock name applies to impure sandstones that contain abundant pre-existing rock fragments together with poorly sorted and rounded mineral grains. This rock contains a fine-grained matrix of grey–green clay and mudstone fragments and more obvious coarse grains of grey, glassy quartz and irregular pink feldspars. Notice the large variation in grain size compared with that of sample 11 and the lack of grain rounding which indicates little 'processing' at the Earth's surface between the sites of weathering and deposition.

Objectives for this Block

When you have completed this Block, you should be able to:

1   Define in your own words, or recognize valid definitions of the terms, concepts and principles listed in Table A (in some way all the ITQs and SAQs in this Block relate to this Objective).

2   Outline the main features of the Solar System and recognize some of the physical and chemical factors that are relevant to the composition, origin and evolution of the planets. (ITQs 1 and 6; SAQ 3)

3   Describe the basis for determining the composition of the Sun's atmosphere from spectral observations and recognize correct statements about this composition. (ITQ 3; SAQs 1 and 2)

4   Recognize examples of nuclear fusion reactions that take place during the lifetimes of stars and account for the main features of the 'cosmic element abundances' curve. (ITQs 2 and 3; SAQ 6)

5   Give reasons for the importance of chondritic meteorites in understanding the composition of the Earth and outline the criteria used in selecting the carbonaceous chondrites as the sub-group that most closely represents the composition of non-volatile matter in the PSN. (ITQ 4; SAQs 3 and 6)

6   Recognize definitions of the terms lithophile, chalcophile and siderophile and use them in relation to the broad distribution of chemical elements in a layered Earth. (ITQs 5 and 7; SAQs 4 and 5)

7   Using data concerning the physical and chemical properties of hypothetical and real planets, make broad deductions about their probable compositions and outline how they may have accreted. (ITQs 5 and 6; SAQ 6)

8   Show how the geochemical classification of the elements is directly related to the fundamental Periodic, bonding and electronic characteristics of each element. (ITQs 8 and 9; SAQ 7)

9   Give a fully reasoned account of why a particular chemical element is likely to be concentrated in the metallic, sulphide-rich, or silicate portions of the layered Earth. (ITQs 7 and 9; SAQ 7)

10   Explain the importance of X-ray diffraction in determining the internal structures of minerals on an atomic scale, and carry out simple calculations involving Bragg's equation. (ITQs 10 and 33)

11   Describe, with reasons, the variations in ionic size of chemical elements and show how the radius ratio in ionic bonding determines the co-ordination states of cations and anions in crystal structures. (ITQs 11, 12 and 13; SAQ 7)

12   Distinguish, with examples, the factors that favour isomorphism, polymorphism, partial ionic substitution and solid solution series. (ITQs 13, 14, 15 and 20; SAQs 8 and 9)

13   Build simple models of silicate polytetrahedra with the Home Experiment Kit and describe the structures of the complex silicate anions. (ITQ 16)

14   Explain, with examples, the relationship between atomic structural properties, physical features (e.g. cleavage and density) and cation content of the five main groups of silicate minerals (Table 8). (ITQs 17, 18 and 19; SAQs 9, 10, 11 and 12)

15   Give a reasoned account of the variations in chemical composition among silicate minerals. (ITQs 17, 20, 24 and 25; SAQs 11 and 12)

16   Account for the variations in relative cation bond strengths in silicate minerals and use them to predict which elements will be most easily freed during chemical breakdown. (ITQs 21 and 27; SAQ 12)

17   Explain how the principles used to study the main groups of silicate minerals may be applied to some non-silicate minerals. (ITQs 15 and 22; SAQ 9)

18   Critically evaluate geochemical data obtained by sampling, averaging and from chemical analysis, and carry out simple statistical calculations using different geochemical data. (ITQs 31 and 32; SAQ 13)

19   Explain the principles used in geochemical analysis by (a) wet chemistry; (b) spectrochemistry; (c) mass spectrometry; (d) X-ray fluorescence; and (e) neutron activation analysis; and evaluate the relative precision, accuracy and sensitivity of each method. (ITQ 33)

20   Identify, with reasons, the mineral content of various igneous, metamorphic and sedimentary rocks, and link this to the major element chemical compositions of such rocks. (ITQs 23, 24, 25 and 30; SAQs 14 and 15)

21   Use the parameters of enthalpy and volume for minerals involved in reactions to deduce the broad physical conditions under which they are most stable. (ITQs 28, 29 and 30; SAQ 15)

22   Account for the different temperatures of crystallization and different resistances to chemical breakdown of the primary silicate minerals, and give examples of their behaviour and fate in magmatic, metamorphic and surface processes. (ITQs 26 and 27; SAQ 17)

23   Indicate a broad understanding of where different major rock types are found in the Earth's crust and relate this to the processes of their origin. (SAQs 16 and 17)

24   Show how, by the influence of the dynamic internal and surface processes of the rock cycle, the Earth's continental crust has become chemically differentiated from, and enriched in fusible components relative to, the mantle. (SAQs 16 and 17)

ITQ answers and comments

ITQ 1 From the relationship $P = \rho g d$ we can deduce that the material at the centre of each planet should be more compressed, and, therefore, that average density should increase *in proportion to radius* if the planets all have the same composition. There is indeed a linear size–density relationship for the Earth, Venus, Mars and the Moon (Figure 53) and, while this *does not prove* conclusively that they have the same composition, our assumption stands up to this test. The odd man out is Mercury; its density is much too large for its size, proving that it cannot have the same composition as the other bodies.

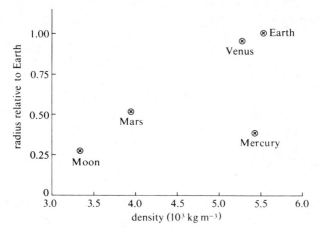

FIGURE 53 Plot of radius against density for the inner planets and the Moon (answer to ITQ 1.)

To anticipate an argument to be developed in full later on, it should be no surprise to you to learn that all the inner planets, except Mercury, are thought to have *broadly* similar compositions, though there are differences in detail. Largely on the evidence of density, coupled with a knowledge of natural element abundances, Mercury is thought to be much more iron-rich and so to have a larger iron core than the other inner planets.

ITQ 2 Figure 7 shows that, beyond iron, the slope of the unbinding energy curve changes in such a way that the average unbinding energy decreases with increasing mass number. Nuclei more massive than iron are more easily unbound (less tightly bound). Moreover, the *difference* between the unbinding energies for a reaction involving the production of more massive atoms (products minus reactants) is now *negative*. This means that energy *must be added* to make nuclei with mass numbers greater than 56, that is, such reactions are *endothermic*[A]. Contrast this with the emission of energy during the fusion of all atoms up to mass number 56.

ITQ 3 The most important change since the Sun formed has been the conversion of hydrogen into helium (equations 2–4). The Sun is not large enough for other nuclear fusion reactions to have occurred. Therefore,

(i) the concentration of hydrogen has decreased;

(ii) the concentration of helium has increased;

(iii) the concentrations of elements other than those that have undergone radioactive decay or fission have remained unchanged.

Notice the implication that the observed relative concentrations of C, N, O, Ne, Mg, Si, S, Fe, etc. in the solar spectrum are unlikely to have changed since the Sun formed.

ITQ 4 (i) The amount of iron in the silicate phase (this will be in the form of Fe_2SiO_4 and other silicates) as a proportion of the total iron content is as follows. The E-group contains no iron silicates and their iron occurs entirely in the reduced metallic, or sulphide forms. At the other end of the scale, the C-group contains no iron metal and these meteorites are completely oxidized. In between, H, L and LL groups contain *c.* 20 per cent, 12 per cent and 6 per cent reduced iron and *c.* 7 per cent, 10 per cent and 14 per cent of iron in silicates. So the order of increasing iron in silicates and of oxidation is: E, H, L, LL, C.

(ii) The total iron content of each group can be read from the parallel diagonal lines because all analyses on one line have the same total iron content. Iron contents are: E 20–34 per cent; H 25–31 per cent; C 22–27 per cent; L 20–24 per cent; LL 18–22 per cent.

Therefore, the order of increasing total iron content is: LL, L, C and E *and* H.

ITQ 5 (i) *Nickel* (Ni) is depleted *c.* 10^2 times in the Earth's crust compared with solar abundances (Figure 11b). But since it is thought to be strongly concentrated in the Earth's core, its total abundance in the Earth *may* be chondritic. (Incidentally, there is evidence, other than that from iron meteorites, that nickel is likely to be concentrated in the core. You will find this in Section 3.)

Potassium (K) is enriched *c.* 50 times in the crust (Figure 11b), and Figure 8 shows that alkali elements are enriched in the crust. In fact, this element is concentrated into partial melts that derive from the mantle and that have formed the crust during the Earth's history.

Magnesium (Mg) is depleted *c.* 10 times in the crust (Figure 11b); Figure 8 shows that magnesium silicates predominate in the mantle. This element is not concentrated in the core, but is more abundant in the mantle than the crust or core.

(ii) As nickel is likely to be concentrated in the Earth's core, magnesium in the mantle and potassium in the crust, this evidence is consistent with the concept of an initially chondritic Earth.

ITQ 6 The approximately linear relationship between size and density for the Earth, Venus, Mars and the Moon (Figure 53) suggests that all these bodies may be chondritic or nearly so (because average density should increase with planetary size, as you learned from the answer to ITQ 1). However, this cannot be true for the dense planet, Mercury, which must contain a greater proportion of heavy elements (probably iron, in view of its abundance in the Solar System—Figure 6) relative to silicates.

ITQ 7 (a) Strongly lithophile: Mg, Th, Ca.

(b) Dominantly chalcophile with some siderophile and/or lithophile tendencies: Cu, Pb.

(c) Dominantly siderophile: Au, Os, Pt.

Although copper and lead are dominantly chalcophile, they are found in both the silicate and metal phases. Calcium, magnesium and the minor element, thorium, are very strongly lithophile and occur in extremely low concentrations in both the metal and sulphide phases. Gold, osmium and platinum show a strong siderophile tendency, much stronger than do nickel and iron! Notice how the alkali metals (Li, Na, K, Rb, Cs) in Table 4 are so strongly lithophile that they are virtually absent from the metal and sulphide phases.

ITQ 8 (a) Progressing from left to right you should recall that s and p electron sub-shells are gradually filled with 2 and 6 electrons respectively until the inert gas configuration is reached. The closer to having a full sub-shell an element is, the easier it is for it to *gain* electrons, thus forming anions. Conversely, the nearer an element is to having empty s and p sub-shells, the easier it is for it to lose electrons and so to form *cations*. That being so, the value of electronegativity—a measure of anion-forming ability—must increase from left to right.

(b) This one is slightly more tricky, but you may have realized that the answer is to do with the increasing number of electron shells around the nuclei of elements in higher Periods (Figure 14). This means that, as you go down the Table, the distance between the nucleus and the most remote p and s sub-shells increases, and so the force of electrostatic attraction becomes weaker between the nucleus (which is positively charged) and the most remote electrons (which are negatively charged). So in Groups VI and VII it is more difficult for high Period elements to attract an electron by electrostatic forces. Thus, bromine is less strongly electronegative than chlorine or fluorine. Similarly, the electrons of the most remote s sub-shell for Groups I and II are more weakly held at higher Periods and so are more easily lost to form cations. Thus, the most 'electropositive' cations are at the bottom left of Table 5 and the most electronegative anions are at the top right.

ITQ 9   Magnesium has the lowest electronegativity and so will tend to form ionic bonds with oxygen (in the ratio 1:1). But you now know (equation 7) that silicon forms complex anions with oxygen which are then linked by cations. The ratio of silicon to oxygen atoms varies quite considerably and need not worry us here (this will be discussed further in Section 3.3). What is clear is that silicate anion complexes will use up much of the remaining oxygen (see Figure 13) and will be bonded by magnesium ions. So, instead of simple oxides of magnesium and oxygen, we obtain *magnesium silicates*. Other elements that are lithophile (with electronegativity less than 1.8 and more easily oxidized than iron, see Table 6) will occur in the magnesium silicate layer—the mantle—of the Earth (Figure 8 and equation 8).

After magnesium silicates have formed there may still be some oxygen remaining: remember Figure 13 gives per cent *by weight* of the different elements and that oxygen has the lightest atoms. Some iron may, therefore, be forced into the lithophile state, as iron oxides or iron silicates. The remaining iron then combines with the available sulphur to form a dominantly chalcophile layer (the outer core, Figure 8) but some iron remains and forms a metallic, siderophile layer (the inner core). Elements that are more siderophile than iron (electronegativity approaching 2.4) will go into the inner core whereas others, with similar electronegativity to iron, will be chalcophile. The inner core–outer core boundary will be controlled by equation 9. (Of course, this assumes that the outer core is an iron–sulphur mixture, which is probable but not certain; this will be discussed in Block 7.)

ITQ 10   The Bragg equation states that:

$$n\lambda = 2d\sin\theta$$

In this case, $\theta = 15.9°$, and $\sin\theta = 0.2740$

$$n = 1$$
$$\lambda = 154\,\text{pm}$$

So $1 \times 154 = 2 \times d \times 0.2740$

from which $d = \dfrac{154}{0.2740 \times 2}$

$$= 281\,\text{pm}$$

From Figure 18 such an olivine contains 35 per cent Mg_2SiO_4 and 65 per cent Fe_2SiO_4.

ITQ 11   (a) In every case where variation in size is shown within a single Group, ionic radius increases down the Group. This must be related to the number of electron shells around the central nucleus: there is at least one additional s and p shell, and in some cases, an additional d sub-shell for each successively higher Period.

(b)   Variations along individual Periods from Groups I to IV are to successively smaller cations because the ionic structure contracts each time an electron is removed to leave a larger net nuclear attraction on each electron that remains. There is a dramatic increase in size where the change from cation-forming to anion-forming elements occurs. Electrons are added to form anions, which decreases the net nuclear attraction on each individual electron and expands the ionic structure.

ITQ 12   Using the ionic radii in Figure 21:
(a) Si^{4+}—O^{2-}, radius ratio = 42/140 = 0.30

(b) Rb^+—O^{2-}, radius ratio = 147/140 = 1.05

(c) Fe^{2+}—O^{2-}, radius ratio = 74/140 = 0.53

(d) Sr^{2+}—O^{2-}, radius ratio = 112/140 = 0.80

(e) B^{3+}—O^{2-}, radius ratio = 23/140 = 0.16

From Table 7, the anticipated co-ordination numbers would be 4 for silicon, 12 for rubidium, 6 for iron, 8 for strontium and 3 for boron.

ITQ 13   The relevant ionic radii are: $Al^{3+} = 51\,\text{pm}$; $O^{2-} = 140\,\text{pm}$; $Si^{4+} = 42\,\text{pm}$; $Fe^{2+} = 74\,\text{pm}$; $Mg^{2+} = 66\,\text{pm}$.

(a)   The radius ratio of $Al^{3+}/O^{2-} = 51/140 = 0.36$, so (from Table 7) it is approaching the upper limit of tetrahedral and the lower limit for octahedral co-ordination, and this explains the appearance of aluminium in both kinds of site.

(b)   There is little chance that Al^{3+} will substitute extensively for either of Mg^{2+} or Fe^{2+} in minerals because the differences in size are 15/51 and 23/51 (29 per cent and 45 per cent) respectively. This exceeds the normal limit for complete ionic substitution of 15 per cent.

(c)   There *cannot* be complete substitution of Al^{3+} for Si^{4+} in minerals because the difference in size (9/42, or 21 per cent) is still more than 15 per cent. Therefore, we must infer that only *partial* ionic substitution occurs. In fact, aluminium is here substituting for silicon in tetrahedral sites; it can do this to a maximum limit of 50 per cent only, because it forms too large an ion for complete tetrahedral site occupancy.

ITQ 14   (a)   Remember that solid solution between two end-members only occurs where there is complete ionic substitution. KCl and KBr form a solid solution series because Cl^- ($r = 181\,\text{pm}$) is very similar in ionic radius to Br^- ($r = 195\,\text{pm}$). Cl and Br can thus replace each other easily in these alkali halides. For NaCl and KCl, replacement of Na^+ ($r = 97\,\text{pm}$) by K^+ ($r = 133\,\text{pm}$) is more difficult because these ions have a much greater difference in radius (> 15 per cent). Therefore you would be right to predict no solid solution between NaCl and KCl; in fact, there is *very limited* ionic substitution.

(b)   The alkali feldspars, like the alkali chlorides, will show only partial ionic substitution between the Na and K end-members (this will be discussed in Section 3.3.4).

ITQ 15   (a)   Assuming a linear polymorphic transition boundary, your sketch should qualitatively resemble that in Figure 54, with aragonite as the high-pressure and low-temperature polymorph of $CaCO_3$, and calcite as the low-pressure and high-temperature polymorph.

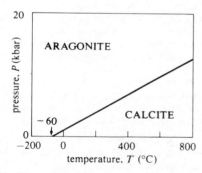

FIGURE 54   The polymorphic transition boundary for calcite and aragonite (answer to ITQ 15).

(b)   Aragonite is stable at higher pressures and therefore is the denser polymorph of $CaCO_3$. (Densities are: calcite, $c.\ 2\,700\,\text{kg m}^{-3}$; aragonite, $c.\ 2\,940\,\text{kg m}^{-3}$.)

(c)   This polymorphic transition must be reconstructive for aragonite to appear at the surface. In fact, it changes to calcite slowly—but not as slowly as diamond to graphite. No aragonite is found in rocks that have been at the surface for more than 100 Ma.

ITQ 16   (a)   The simple tetrahedron has a unit formula SiO_4 and has a net charge of -4, i.e. $(SiO_4)^{4-}$.

(b)   Each tetrahedron in a row shares two oxygen atoms with adjacent tetrahedra, so the simplest unit formula is SiO_3, with a net charge of -2, i.e. $(SiO_3)^{2-}$.

(c)   The double row is joined by an additional oxygen atom being shared in every *second* tetrahedron. Thus, alternate tetrahedra share two and three oxygens and so, for every 2 silicons there are $3 + 2.5$

oxygens. The simplest unit formula with whole numbers of oxygen and silicon atoms is therefore Si_4O_{11}, for which there is a net charge of $16 - 22 = -6$, i.e. $(Si_4O_{11})^{6-}$.

(d) The two-dimensional array is characterized by every tetrahedron sharing three of its oxygens with adjacent members of the sheet. Only the fourth oxygen projects above and below the sheet and is available for other forms of bonding. In the sheet there are 2.5 oxygens for every silicon, giving a simplest unit formula of Si_2O_5, with a net charge of $8 - 10 = -2$, i.e. $(Si_2O_5)^{2-}$.

(e) In a three-dimensional array, all the oxygens must be shared equally between two silicons, so there are two oxygens for each silicon. Thus, the simplest unit formula reduces to SiO_2, which is electrically neutral, and is the formula for quartz.

ITQ 17 From Table 8, the following can be deduced:

Mineral	Cations	Silicon atoms	Cation:silicon ratio
Olivines	2	1	2:1
Pyroxenes	2	2	1:1
Amphiboles	7	8	7:8

So the number of cations per silicon atom decreases down the table; olivines therefore have the lowest proportion of silicon and chain silicates the highest. In fact, most amphiboles are more complicated than tremolite and contain aluminium substituting for silicon. So pyroxenes usually have a greater proportion of silicon than amphiboles, contrary to what you may have predicted.

ITQ 18 If there are six Mg^{2+} ions for every eight Si^{4+} ions, then, remembering that the repeat unit in sheet silicates is $[Si_4O_{10}(OH)_2]^{6-}$, which has *four* silicon atoms, the mineral shown in Figure 35a must have the formula: $Mg_3Si_4O_{10}(OH)_2$. This is the mineral *talc*.

ITQ 19 If one in four silicons (Si^{4+}) in quartz is replaced by aluminium (Al^{3+}) then there will be a charge deficiency of one positive charge per four SiO_2 groups, i.e. $(AlSi_3O_8)^-$. This can be satisfied by the incorporation of one univalent cation into the structure. The most common of such ions are Na^+ and K^+; the addition of these ions to the aluminosilicate anionic structure (above) results in the formation of the *alkali feldspar* mineral group.

ITQ 20 (a) There has to be more replacement of silicon by aluminium in calcium feldspar ($CaAl_2Si_2O_8$) than in sodium feldspar ($NaAlSi_3O_8$) to preserve electrical neutrality, and this is why the ratio of $Al_2O_3:SiO_2$ is higher in the former.

(b) Sodium and potassium feldspars each have the same number of molecules of alkali oxide, SiO_2 and Al_2O_3. But potassium is heavier than sodium and so the K_2O content of orthoclase is higher than the Na_2O content of albite. In turn, this means that the weight percentages of Al_2O_3 and SiO_2 are lower in orthoclase than in albite.

ITQ 21 See Table 17 (a completed form of Table 10).

TABLE 17

Cation	Co-ordination state	Relative bond strength
Si^{4+}	Tetrahedral (4-fold)	1
Al^{3+}	Tetrahedral (4-fold)	3/4
Al^{3+}	Octahedral (6-fold)	1/2
Mg^{2+}	Octahedral (6-fold)	1/3
Fe^{2+}	Octahedral (6-fold)	1/3
Ca^{2+}	Cubic (8-fold)	1/4
Na^+	Cubic (8-fold)	1/8
K^+	12-fold	1/12

ITQ 13 explains the 'dual' behaviour of aluminium. Calcium usually enters 8-fold, rather than 6-fold, co-ordination sites; the same is true for sodium which forms ions that are only slightly larger. (If you used 6-fold co-ordination for these two elements you would obtain about the same order of relative bond strengths.)

ITQ 22 The ionic radii of Ca^{2+} and F^- are 99 pm and 136 pm respectively, giving a radius ratio of 0.73. Although you would be right to think that this implies octahedral co-ordination, in fact, this is just at the limit where cubic co-ordination becomes necessary (Table 7) and each calcium atom is surrounded by eight fluorine atoms. Since there are no additional atoms to complicate matters, the external morphology, like the internal crystal structure, should be cubic (Figure 38).

ITQ 23 See Table 18 (a completed form of Table 12).

TABLE 18 Silicate minerals in major rock types

Rock type	Olivine	Garnet	Pyroxene or amphibole	Mica	Feldspar	Quartz	Clay minerals
IGNEOUS							
Granite	—	—	✓	*	✓	✓	—
Gabbro	*	—	✓	—	✓	—	—
Peridotite	✓	—	*	—	*	—	—
SEDIMENTARY							
Pure sandstone	—	—	—	—	—	✓	—
Greywacke	—	—	—	—	✓	✓	*
Mudstone	—	—	—	—	*	*	✓
METAMORPHIC							
Mica schist	—	*	—	✓	✓	*	*
Amphibolite gneiss	—	—	✓	—	✓	—	—
Granite gneiss	—	*	—	✓	✓	*	—

You should have managed to get most of the items ticked in this table. The asterisks are more tricky—we have included them for completeness—and they indicate low or variable abundances of minerals in the rock types indicated.

ITQ 24 (a) All the minerals present in both rocks, except iron oxide, are silicates—but as you can see from Table 9, *they contain different amounts of* SiO_2. The granite analysis contains contributions of SiO_2 from quartz (25 per cent modal) and from K-feldspars (55 per cent modal) which are also relatively rich in SiO_2 (c. 65 per cent by weight in Table 9). On the other hand, the basalt contains no quartz, but contributions from silica-poor minerals such as olivine and pyroxene.

(b) The iron- and magnesium-bearing minerals in granite are few: biotite mica (9 per cent modal) and iron oxides (1 per cent modal). Basalt contains much greater amounts of such minerals and includes pyroxene (30 per cent modal), olivine (10 per cent modal) and iron oxides (5 per cent modal). Reference to Table 9 shows that pyroxene and olivine are relatively richer in $FeO + MgO$ than in biotite mica. So basalt contains a greater proportion of ferromagnesian minerals than granite; its ferromagnesian minerals also contain more Fe and Mg than those of granite.

(c) The CaO content of basalt (11.1 per cent by weight) is made up of contributions from pyroxene (30 per cent modal, which contains c. 15 per cent CaO by weight) and from plagioclase (50 per cent modal, which contains c. 12 per cent CaO by weight). The small amount of CaO in the granite analysis is contributed entirely by plagioclase feldspar (10 per cent modal).

As we noted in Section 3.3.5, the minerals contained in an igneous rock are determined by the chemical composition of the magma that crystallized to form them—in this ITQ you have been considering the relationship between minerals and rock compositions the other way round.

ITQ 25 (a) (i) Oxides that increase in concentration all the way to granite are: SiO_2 and K_2O.

(ii) Those that decrease are: FeO, MnO and MgO.

(iii) The remainder, that rise and then fall are: TiO_2, Al_2O_3, Fe_2O_3, CaO, Na_2O, P_2O_5 and H_2O.

(b) (i) Quartz and K-feldspar increase from basic to acid rocks so the amount of silica and potassium must also increase.

(ii) The content of ferromagnesian minerals decreases from basic to acid rocks. Manganese is similar geochemically to iron (see Table 5) and occurs as a minor element in all iron-bearing minerals.

(iii) This is a bit more difficult, but can be solved using Figure 41. The increase and then decrease in Al_2O_3 and Na_2O is explained almost entirely by the abundance of plagioclase feldspar in intermediate rocks compared with acid and basic rocks (Figure 41 and Table 9). CaO and the minor elements TiO_2 and P_2O_5 are present in pyroxene, amphibole and other minor mineral phases such as Fe–Ti oxides and apatite (Table 11), which are most abundant in basic and intermediate rocks. Water is present (as OH^-) in micas and amphiboles (as you may remember from SAQ 10), and these are more abundant in intermediate rocks than in acid or basic rocks (Figure 41). (The occurrence of a maximum Fe_2O_3 concentration in intermediate rocks is probably related to the abundance of magnetite, Fe_3O_4. Magnetite has one Fe^{3+} ion for every Fe^{2+} ion in its structure. Notice, however, that *total* iron (FeO + Fe_2O_3) varies smoothly right across Table 14.)

ITQ 26 (a) Table 15 shows that magnesium-, iron- and calcium-bearing minerals crystallize before sodium- and potassium-bearing minerals. Taking the lists of major minerals in granite and basalt from Table 13a, the order (with decreasing temperature) will be:

granite plagioclase feldspar, biotite mica, K-feldspar, quartz.

basalt olivine, pyroxene and plagioclase feldspar, K-feldspar.

(b) As you can see from Table 15, most of the minerals in basalt crystallize at higher temperatures (1 500–1 200 °C) than most of those in granite (1 200–800 °C).

ITQ 27 Quartz is the most resistant mineral to chemical weathering and olivine the least. It follows that the basalt will be much more easily decomposed than the granite. The order of increasing resistance of silicate minerals to chemical weathering is the same as the order of their crystallization from igneous magmas (i.e. towards increasing structural complexity, Table 15).

The products of basalt weathering will be soluble Mg^{2+}, Ca^{2+} and Na^+ from olivines, pyroxenes and feldspars, together with soluble silica, some clay mineral particles from the feldspar and insoluble iron-rich precipitates. Granite weathering will also produce soluble K^+, Na^+ and Ca^{2+}, together with large amounts of clay minerals (from feldspar and mica) and unaltered quartz.

ITQ 28 The correct answers are (a) and (c). Endothermic reactions have a positive ΔH (Figure 43) and so require an energy input which is supplied by increasing temperature. The lower-volume (higher-density) assemblage is also favoured by conditions of increased pressure. You may remember that aragonite is the denser polymorph of $CaCO_3$ (ITQ 15) and is therefore favoured by conditions of higher pressure.

ITQ 29 (a) The left-to-right reaction must be endothermic, because it requires an input of thermal energy for it to proceed during metamorphism. In fact, ΔH is $+89\,kJ\,mol^{-1}$ for this reaction.

(b) There is an increase in volume: $CaCO_3$ and $CaSiO_3$ are both solids and, while SiO_2 is a solid, CO_2 is a gas that occupies more space than the equivalent amount of solid. So ΔV must be positive.

(c) ΔV and ΔH are both positive for this reaction, so there will be a boundary curve with a positive gradient on a pressure–temperature diagram. The relevant diagram is shown in Figure 55; the wollastonite + carbon dioxide assemblage is favoured by high temperatures and low pressures. High pressures do not allow CO_2 to expand and escape and so drive the reaction back to the left-hand side. (The boundary is curved, rather than being a straight line because ΔV and ΔH vary differently with changes in pressures and temperatures.)

ITQ 30 (a) For each boundary curve, the polymorph on the high-temperature side has the highest enthalpy and that on the high-pressure side has the lowest volume (highest density). Thus, the order of increasing enthalpy is kyanite, andalusite, sillimanite and the order of increasing volume is kyanite, sillimanite, andalusite. (In fact, the densities are kyanite = $3.6 \times 10^3\,kg\,m^{-3}$; sillimanite = $3.25 \times 10^3\,kg\,m^{-3}$; andalusite = $3.15 \times 10^3\,kg\,m^{-3}$.)

(b) Andalusite, the low-pressure, medium-temperature polymorph, will occur most commonly in contact metamorphism (compare Figures 46, 48 and 49 and note that the spots in spotted slates are often andalusite crystals). The boundary between the kyanite and sillimanite fields runs through the high-temperature side of the amphibolite facies field (Figure 48) and kyanite will be most common in this facies.

ITQ 31 (a)

$$\text{Mean } \bar{x} = \frac{19 + 20 + 17 + 21}{4}$$

$$= 19.25$$

$$\text{Standard deviation, } s = \sqrt{\frac{0.25^2 + 0.75^2 + 2.25^2+ 1.75^2}{4}}$$

$$= \sqrt{\frac{8.75}{4}} \qquad = \sqrt{2.19}$$

$$= 1.48 \text{ p.p.m.}$$

$$\text{Coefficient of variation, } C_v = \frac{100 \times 1.48}{19.25}$$

$$= 7.7 \text{ per cent}$$

(b) This information allows us to specify how closely the data approximate to the mean. 68 per cent of the readings will have values of 19.25 ± 1.48 p.p.m., but the 95 per cent confidence limit ($\bar{x} \pm 2s$) will contain values ranging from 19.25 ± 2.96, that is, from 16.3 to 22.2 p.p.m.

(c) In view of the statement that C_v for trace element analyses often reaches values near 1 per cent, the precision of these analyses ($C_v = 7.7$ per cent) is on the poor side, but it may well have been adequate for the problem involved.

ITQ 32 CaO = $20.8/200 \times 100 = 10.4$ per cent. Basalts (Table 13) or gabbros (Table 14) have this order of CaO content. (Of course, in a basalt, calcium will be present in the combined form with silicate-forming minerals such as pyroxene and plagioclase feldspar.)

ITQ 33

$$n\lambda = 2d \sin \theta$$

$$713 \text{ pm} = 874 \sin \theta$$

$$\sin \theta = 0.8158$$

Hence $\theta = 54° 40'$

This is the angle at which the analysing crystal must be set relative to the direction of the incident X-ray beam in Figure 52.

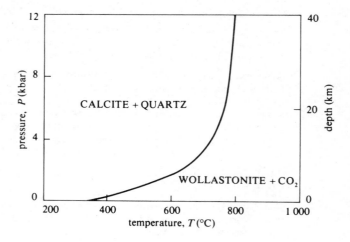

FIGURE 55 Boundary curve for the reaction in ITQ 29.

SAQ answers and comments

SAQ 1   The line intensities, and hence the concentrations, for both K and Rb, decrease in the order mica–granite–basalt. Mica is the only sample that records caesium. The order of decreasing Na content is granite–basalt–mica. Notice that mica, which contains high concentrations of Cs, K and Rb, has the lowest Na content. Granite is relatively rich in K, Rb and Na. We shall return to these observations in Section 4.

SAQ 2   (a)   Reference to Figure 3 shows that these are the two frequencies of the prominent *hydrogen* lines.

(b)   Reference to Figure 6 shows that (i) is *valid*: hydrogen is the most abundant cosmic element and there is a general decrease in abundance for elements of higher atomic number, and that (ii) is *invalid*: elements with even atomic numbers are *more* abundant than those with odd atomic numbers.

SAQ 3   (i) B, D, G; (ii) B, E, F; (iii) C; (iv) A.

Check back to Section 2.1 if you had any of these wrong.

SAQ 4   (i) A; (ii) B; (iii) C; (iv) A; (v) C; (vi) B.

Check back to Section 2.3 if you had any difficulty. Bear in mind that the *explanation* of these correlations is yet to be developed; we shall do this in Section 3.

SAQ 5   Mass of gold in mantle $= \dfrac{0.005}{10^6} \times 4 \times 10^{24} = 2 \times 10^{16}\,\text{kg}$.

Mass of gold in core $= \dfrac{1.5}{10^6} \times 2 \times 10^{24} = 3 \times 10^{18}\,\text{kg}$.

Total mass of gold in the Earth's mantle and core $= 3.02 \times 10^{18}\,\text{kg}$ and the *proportion* of the Earth's gold that is in the core $= 3 \times 10^{18}/3.02 \times 10^{18} = 99.3$ per cent.

SAQ 6   (a)   True. Refer to Section 2.3.1.

(b)   False. Elements with mass numbers that are multiples of four are most abundant in this range; these result from *helium* nucleus additions, not hydrogen (which can add only 1 or 2 mass units). (Section 2.2.2.)

(c)   True. Refer to Figure 6: the composition of the PSN is recorded approximately in solar spectra and in the cosmic element abundance curve.

(d)   False. Helium burning releases energy, whereas a supernova results from reactions that seek to withdraw energy from the star. Such endothermic reactions dominate the last stages of a large star's life and involve the production of elements heavier than iron (Figure 7 and Section 2.2.2).

(e)   True. For short-lived radioisotopes to have been incorporated into the Solar System, they must have been produced a short time before the Solar System formed and a supernova is the only means of releasing such isotopes from pre-existing stars—the Sun cannot manufacture elements heavier than helium. (Section 2.2.2.)

(f)   True. For iron to be separated from silicates, so forming a core and mantle, melting temperatures are needed and these two processes of heating are the most likely (Section 2.3).

(g)   False. The basis of the chondritic Earth model is that the whole Earth is chondritic, not just the crust (Section 2.3.2).

SAQ 7   (a)   It is generally the case that the more easily an element gives up an electron to form a cation, that is, the less electronegative it is, the more easily it will become oxidized (see Table 6 and Section 3.1.2). So the order of ease of oxidation will be B, D, Fe, C, A.

(b)   Radius ratios relative to oxygen (140 pm) are: A 0.99; B 0.76; C 0.54; D 0.33; Fe 0.53.

Thus, cation D would be in *tetrahedral* co-ordination, with C and Fe in *octahedral* co-ordination, B in *cubic* co-ordination and A in *12-fold* co-ordination with oxygen (see Table 7). Of course, it is unlikely

that element A, and also probably C ever would form ionic bonds with oxygen because of their high electronegativities (see (c)).

(c)   Elements B and D will be lithophile, element C will be chalcophile and element A siderophile on the basis of electronegativities. The ionic radius and charge of D are such that it would *form* frameworks (as silicon does in silicates, Section 3.2.1), whereas the larger ion of lower charge, B, will *link* frameworks (it has the characteristics of alkali elements—see Table 5 and Figure 21). So the answers are: (i) D; (ii) B; (iii) C; (iv) A.

(d)   It is a simple step from (c) to infer that A will occur in the inner core, C will occur in the outer core, and B and D will occur in the crust and mantle.

SAQ 8   Fluoride (F^-) has an ionic radius of 136 pm (Figure 21). Since this differs from $(OH)^-$ ($r = 140\,\text{pm}$) by less than 15 per cent, F^- *can* replace $(OH)^-$ in a solid solution series (and, incidentally, strengthen the calcium phosphate structure in teeth) (Section 3.2.2). Chloride (Cl^-), however ($r = 181\,\text{pm}$), has a radius that is more than 15 per cent greater than that of $(OH)^-$, and so it *cannot* be incorporated so easily into the chemical compounds present in teeth and bone.

SAQ 9   (a)   (i) As in iron- and magnesium-bearing olivines, there is complete ionic substitution and therefore a solid solution series between siderite ($FeCO_3$) and magnesite ($MgCO_3$) because the ionic radii of iron and magnesium differ by less than 15 per cent. Also as in olivines, the calcium ion is too large for much ionic substitution of Fe^{2+} or Mg^{2+} by Ca^{2+} to take place and so there is no solid solution series between $CaCO_3$ and either $FeCO_3$ or $MgCO_3$. (Remember from Section 3.2.2 that isomorphous substances do not necessarily form solid solutions.)

(ii) You would not have been able to predict this from the information available. The ionic radii of Na^+ (97 pm) and Ca^{2+} (99 pm) are sufficiently close for solid solution to be possible. However, we do not know whether the sizes of the anions NO_3^- and CO_3^{2-} are sufficiently similar. In fact, there is a considerable size difference and the two substances do not form a solid solution series.

(b)   If you look at the two chemical formulae, you will see that this question is asking whether there is complete ionic substitution between Mg^{2+} and Al^{3+} in octahedral sites. Their ionic radii (66 pm and 51 pm) differ by nearly 30 per cent and so there is no solid solution series between biotite and muscovite.

SAQ 10   (a)   *Amphiboles* are double-chain silicates whose structures contain hexagonal holes which provide sites for *structurally bonded* hydroxyl $(OH)^-$ groups (Figures 31, 32 and 34). Notice that in the chemical formulae of amphiboles (Table 8) the negatively charged hydroxyl groups balance the structure electrically; this indicates structurally bonded water.

(b)   The situation is similar in sheet silicates, of which white mica, *muscovite*, is an example (Figures 33–35).

(c)   *Kaolinite*, a clay mineral, is another sheet silicate but with a different internal structure from that of muscovite (Figure 36b). Octahedrally co-ordinated cations link these sheets to alternating layers of *hydroxyl* complexes (hydroxyl groups also occur in hexagonal holes in the sheets) and, therefore, again the water is structurally bonded. Inter-layer water may also occur in the form of free water molecules in the kaolinite structure, but these are less abundant than in montmorillonite.

(d)   *Montmorillonite*, another clay mineral, comprises sandwich-like sheets (similar to those in mica) which contain hydroxyl groups in the hexagonal holes. But by far the greater amount of water usually associated with montmorillonite occurs between the sandwiches (Figure 36a) as *unbonded water molecules* which, if present, cause the clay structure to swell to a low-density mass.

SAQ 11   Minerals that contain iron and/or magnesium in significant amounts, are dark in colour and relatively dense, are *ferromagnesian*: (b), (d), (e) and (f). *Salic* minerals are usually more silica-rich; apart from quartz, they contain aluminium (in octahedral sites); they are light in colour and of relatively low density: (a), (c), (g) and (h). (We hope you found this question pretty straightforward, but see Section 3.3 and Tables 8 and 9 if you had any difficulty.)

SAQ 12 (a) The atomic structure of olivine (Figure 28) is more *densely packed* with silicate tetrahedra than that of micas (Figures 33–35), which have holes within the sheets, often occupied by light water molecules, and spaces between the sheets, partly occupied by large potassium atoms. The other important factor is that olivines contain a significant amount of *iron*, which has heavy atoms, whereas micas contain *lighter molecules*, such as hydroxyl groups. $(OH)^-$ groups (molar mass = 17) occupy more space in micas than Fe^{2+} ions (atomic mass = 56) do in olivines.

(b) Remember that amphiboles are double-chain silicates whereas micas are sheet silicates; both have well-developed cleavages (e.g. Figure 32c for amphiboles). But the mica sheets cleave more easily because the bonds holding the sheets together are weaker than those holding the double chains together in amphiboles. The relative bond strength of sheet-linking potassium ions in micas is 1/12 (Table 10) whereas the chain-linking cations (Mg^{2+}, Fe^{2+}, Ca^{2+}, Na^+, etc.) in amphiboles have relative bond strengths ranging from 1/3 to 1/8.

(c) A plagioclase feldspar in the middle of the solid solution will have a chemical formula half-way between $NaAlSi_3O_8$ (albite) and $CaAl_2Si_2O_8$ (anorthite). From Table 10, you should realize that divalent calcium ions are held in the framework more strongly than univalent sodium ions (relative bond strengths are 1/4 and 1/8 respectively). So chemical weathering of this feldspar is likely to break sodium–oxygen bonds first, releasing sodium into solution, then calcium–oxygen bonds, releasing calcium. This leaves the electrically charged Al—Si—O framework, which is resistant to chemical attack, as small (clay-sized) particles. As you will find in Section 4, these particles become structurally reorganized into clay minerals; this involves the breaking of some Al—O bonds. The most extreme conditions of chemical weathering could break apart the entire structure, yielding still smaller molecules.

SAQ 13 (a) Since coverage of the whole of Cornwall is required, a *systematic* sampling plan (ii) would be used. However, to detect the areas *richest* in tin, only *precise* measurements would be needed (iv).

(b) Since we are not concerned with distribution in a given area, it is better to collect samples as widely and *randomly* (i) as possible. Again, only *precise* measurements (iv) are needed since we are not seeking absolute values.

(c) *Systematic* sampling of the particular area (ii). Since we need an average composition, *accurate* analysis would be necessary (iii).

(d) *Random* samples of mica (i) would be used from a great variety of rock types and, since we are trying to establish averages, the analyses would need to be as *accurate* as possible (iii).

SAQ 14 (a) To judge from its mineralogy, this rock is intermediate between a quartz-rich granite and a feldspathic diorite. It probably lies in the *granodiorite* area with a composition similar to that in column 2 of Table 14.

(b) By comparison with Table 14, this can only be an olivine-rich peridotite.

(c) The high CO_2 content of this rock is diagnostic of carbonate sediments (Table 16). This indicates a limestone, and the mixture of MgO and CaO indicates that the single mineral is not calcite ($CaCO_3$). This is a partly dolomite limestone, $CaMg(CO_3)_2$. It contains the mineral dolomite, which is isomorphous with calcite.

(d) This is a sandstone which, with 80 per cent quartz and 20 per cent alkali feldspar, should have a chemical composition of over 90 per cent SiO_2, with small amounts of Al_2O_3 and alkali oxides. Therefore it is a sandstone of greater purity than analysis 2 of Table 16 but the feldspar content indicates incomplete chemical breakdown to clay minerals during weathering.

SAQ 15 (a) Calcite is most stable at low pressure and high temperature; aragonite is most stable at low temperature and high pressure (see also the answer to ITQ 15). Therefore,

(i) *calcite* has the higher enthalpy—aragonite to calcite reactions are endothermic and have a positive enthalpy of reaction (ΔH) and vice versa;

(ii) *calcite* also has the higher volume (lower density)—aragonite to calcite reactions lead to an increase in volume (positive ΔV) and vice versa. (See also Figure 45.)

(b) (i) Basalt is metamorphosed to *amphibolite* at high pressures and temperatures. (Such rocks are rich in amphiboles and feldspars; they are characteristic of the amphibolite facies of metamorphism.)

(ii) Sandstones recrystallize under metamorphism to form *quartzites*. Hornfels is formed by contact metamorphism, usually of clay-rich sediments, whereas marbles result from the metamorphism of limestone.

SAQ 16 (c) and (d) are true—see Sections 4.1 for (d) and 4.3 for (c); (a) and (b) are false.

(a) Figures 46 and 48 show that blueschists are formed at the high pressures of burial metamorphism rather than the high temperatures of contact metamorphism.

(b) is totally wrong—the process of crust formation depends on the Earth's *internal* heat sources that distil crust-forming materials from the mantle (Figure 50 and Sections 4.1 and 4.4). Solar energy drives the surface/sedimentary part of the cycle and cannot influence processes more than a few metres below the ground surface.

SAQ 17 (a) (i) Ferromagnesian minerals are not common in many sedimentary rocks because some iron and most magnesium is released into *solution* by chemical weathering: they are carried as soluble materials and accumulate in ocean waters. Therefore, they do not contribute significantly to the insoluble products of surface processes, sandstones and mudstones, except in that clay minerals, such as montmorillonite (Figure 36a), may resorb cations into interlayer structural sites. Some iron accumulates as relatively insoluble oxides near the weathering site but even this iron has been released from silicates.

(ii) The *rate* of chemical weathering and the *time* between weathering and the formation of new sedimentary rocks are important factors determining the extent to which feldspar breaks down chemically into clay minerals. Therefore, a sand deposit that accumulates near to a primary crystalline rock outcrop, undergoing rapid physical weathering, may contain feldspar, whereas another similar deposit forming in a location remote from the same site may contain clay minerals.

(iii) Metamorphism is a rebuilding process for silicate mineral structures. Therefore, a clay-rich sediment that contains some iron and/or magnesium is ideally suited to produce ferromagnesian minerals (chlorite, biotite, amphibole, etc.) during metamorphism.

(iv) Garnet is a mineral of high density and low volume (Section 3.3.1), which will replace low-density minerals during high-pressure metamorphism.

(b) Look back to Figures 42 and 50. Sodium will be enriched in basic magmas that originate by partial melting of mantle peridotite beneath a spreading ridge. It will accumulate in basaltic ocean crust in feldspar crystals and (assuming that it is not leached by seawater) will be subducted with that crust at a destructive margin. It will also preferentially enter the magmas of intermediate to acid composition that rise from within (and above) the subduction zone. (Note the progressive enrichment of Na_2O from peridotite to granodiorite in Table 14.) It may then crystallize in the feldspars of either 'granitic' intrusions or andesite lavas, from whence it may ultimately be subject to release into the surface environment by weathering and erosion. In the products of chemical weathering sodium is soluble and so it will reach the oceans. It will then either become adsorbed by clay minerals, forming mudstones, etc., which may be metamorphosed, thus recycling sodium through the internal part of the rock cycle, or form salt deposits through the intense evaporation of seawater in an enclosed basin.

Acknowledgements

Grateful acknowledgement is made to the following for permission to reproduce material in this Block:

Figure 4 from H. E. White (1973) *Modern College Physics*, Van Nostrand Reinhold; *Figure 5* Hale Observatories; *Figure 6* based on K. H. Wedepohl (1970) *Geochemistry*, Holt, Rinehart and Winston, reproduced by permission of the author; *Figure 9* from K. Keil, 'Classification of chondrites' in K. H. Wedepohl (ed.) (1974) *Handbook of Geochemistry*, Springer-Verlag; *Figure 10* J. A. Wood in B. M. Middlehurst and G. P. Kuiper (1963) *The Moon, Meteorites and Comets*, University of Chicago Press; *Figure 12* from S. R. Taylor and L. H. Ahrens, 'Spectrochemical analysis' in A. A. Smales and L. R. Wager (eds.) (1960) *Methods in Geochemistry*, Wiley Interscience; *Figures 30, 32 and 35*, J. Lameyre (1975) *Roches et Minéraux—Les Matériaux*, Editions Doin, reproduced by permission of the author; *Figure 52a* Courtesy Philips Analytical Equipment. The text quotation on page 82 is from B. Mason (1966) *Principles of Geochemistry,* 3rd edn., John Wiley & Sons Inc.

The Course Team acknowledges with gratitude the following external assessors:

Dr S. W. Richardson, Department of Geology and Mineralogy, University of Oxford (Course Assessor); Dr P. A. Floyd, Department of Geology, University of Keele (Block 1 Assessor); Dr P. M. Martin, Department of Physics, University of Essex.

Most of the Figures in this Block were drawn by Hugh Bevan, Joanne Clark and Mark Kesby.

The Earth: Structure, Composition and Evolution

TABLE 5 Periodic Table of the elements, showing geochemical tendencies. Electronegativities appear below element symbols. A heavy line surrounds all elements that show either lithophile or atmophile affinities in the Earth.

GROUP / PERIOD	I	II	TRANSITION ELEMENTS (+ LANTHANIDES + ACTINIDES)										III	IV	V	VI	VII	INERT GASES
1	H 2.1																	He —
2	Li 1.0	Be 1.5											B 2.0	C 2.5	N 3.0	O 3.5	F 4.0	Ne —
3	Na 0.9	Mg 1.2											Al 1.5	Si 1.8	P 2.1	S 2.5	Cl 3.0	Ar —
4	K 0.8	Ca 1.0	Sc 1.3	Ti 1.5	V 1.6	Cr 1.6	Mn 1.5	Fe 1.8	Co 1.8	Ni 1.8	Cu 1.9	Zn 1.6	Ga 1.6	Ge 1.8	As 2.0	Se 2.4	Br 2.8	Kr —
5	Rb 0.8	Sr 1.0	Y 1.2	Zr 1.4	Nb 1.6	Mo 1.8	Tc 1.9	Ru 2.2	Rh 2.2	Pd 2.2	Ag 1.9	Cd 1.7	In 1.7	Sn 1.8	Sb 1.9	Te 2.1	I 2.5	Xe —
6	Cs 0.7	Ba 0.9	La-Lu 1.1-1.2	Hf 1.3	Ta 1.5	W 1.7	Re 1.9	Os 2.2	Ir 2.2	Pt 2.2	Au 2.4	Hg 1.9	Tl 1.8	Pb 1.8	Bi 1.9	Po 2.0	At 2.2	Rn —
7	Fr 0.7	Ra 0.9	Ac 1.1	Th 1.3	Pa 1.5	U 1.7	Np-Lr 1.3											

ATMOPHILE LITHOPHILE CHALCOPHILE SIDEROPHILE

FIGURE 21 Some ionic radii (in picometres): the elements are arranged in a much-reduced form of the Periodic Table (see Table 5). Circles are drawn to the same scale and indicate relative ionic sizes. Note that OH^- is shown in a convenient space but it does not really belong to Group VI.

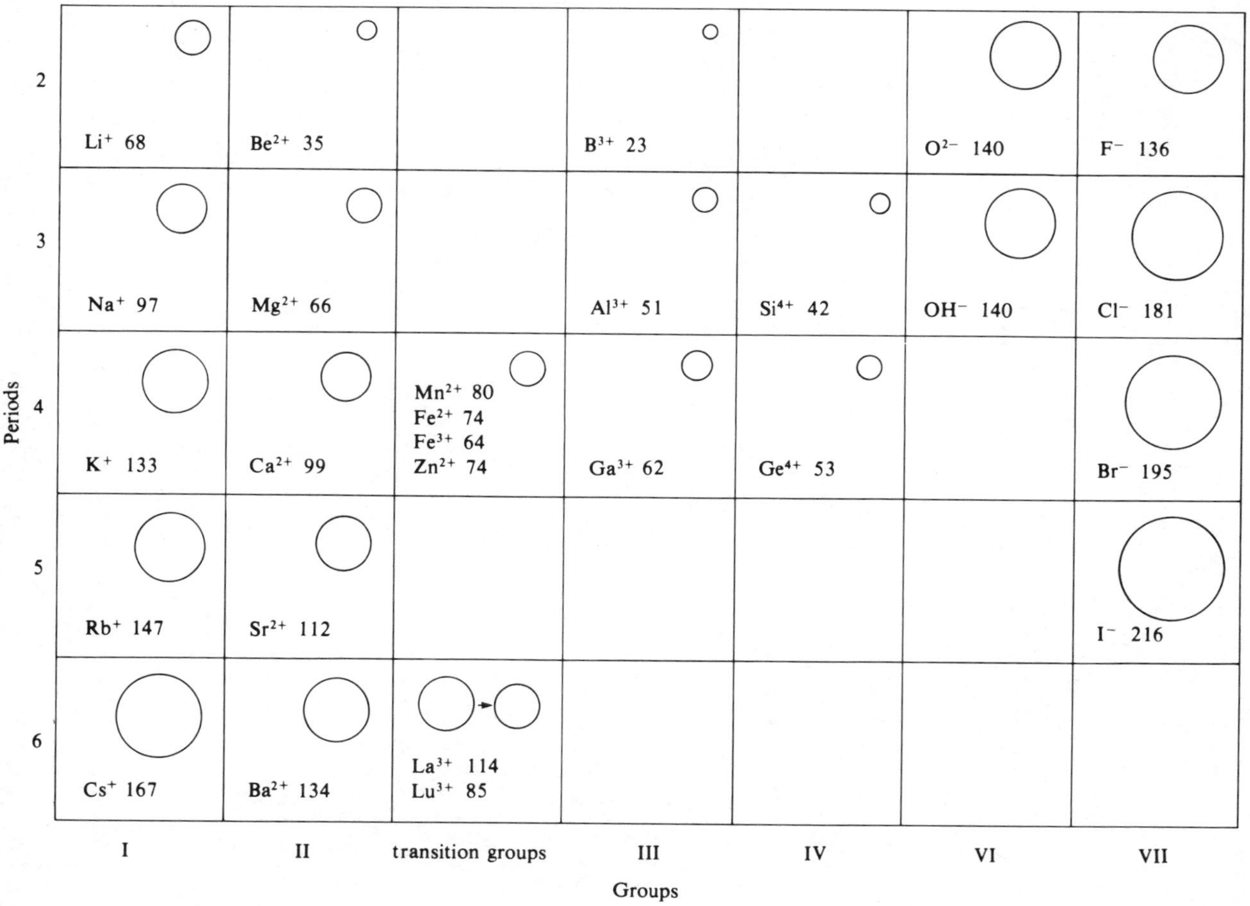